D1226458

DA CAPO PRESS SERIES IN

# ARCHITECTURE AND DECORATIVE ART

*General Editor:* ADOLF K. PLACZEK

*Avery Librarian, Columbia University*

VOLUME 13

# THE ORIGINS OF CAST IRON ARCHITECTURE IN AMERICA

# THE ORIGINS OF CAST IRON ARCHITECTURE IN AMERICA

Including

## ILLUSTRATIONS OF
## IRON ARCHITECTURE MADE BY
## THE ARCHITECTURAL IRON WORKS
## OF THE CITY OF NEW YORK

DANIEL D. BADGER, PRESIDENT

And

## CAST IRON BUILDINGS:
## THEIR CONSTRUCTION AND ADVANTAGES

BY JAMES BOGARDUS

New Introduction by W. KNIGHT STURGES

DA CAPO PRESS • NEW YORK • 1970

A Da Capo Press Reprint Edition

*Cast Iron Buildings: Their Construction and Advantages,*
was originally published in New York in 1856.
*Illustrations of Iron Architecture,*
also published in New York, appeared in 1865.
This Da Capo Press edition is an unabridged
republication of both works.

Library of Congress Catalog Card Number 68-25760
SBN 306-71039-0

© 1970 by Da Capo Press
A Division of Plenum Publishing Corporation
227 West 17th Street
New York, N.Y. 10011
Printed and bound in Japan

# ILLUSTRATIONS
# OF IRON ARCHITECTURE

# INTRODUCTION

———•———

Ignored or considered beneath the notice of serious criticism until a generation ago, cast iron construction is recognized today not only as the precursor of the steel framed skyscraper, but also as an early example of prefabrication and modular design. In the pages which follow we shall see how standardized the building process could become in an earlier day and how, a little over a hundred years ago, the twentieth-century dream of prefabrication was momentarily realized.

*Illustrations of Iron Architecture* is basically nothing more than the catalog of an iron foundry. However, this catalog is both very complete and superbly illustrated. The lithographers, Sarony and Major, made of this book more than a mere catalog. It is rather a compendium of the work of the period and for us today an important source book of cast iron construction in the United States.

Of Daniel D. Badger, the president of the Architectural Iron Works, we know comparatively little. In the foreword it is stated that he came to New York from Boston. He is listed in the New York City Directory for 1849 as a manufacturer of iron shutters, and some years later as an ironmonger. In the Directory for 1865, the year of publication of our book, he is styled, simply, "President," and he retains this title until 1877, when his name disappears from the volume.

From these rudimentary facts we may infer that we are looking at the output of no more than fifteen years, four of them Civil War years. But by the date of publication, Badger's firm had produced more than three hundred buildings in New York City alone, to say nothing of Philadelphia, where forty are listed, or Charleston, South Carolina, which could boast fourteen. Badger's buildings also had been exported to distant parts of the world—Halifax, Havana, Alexandria (Egypt), and Rio de Janiero, to name but a few.

It should be borne in mind, moreover, that Badger's concern was only one of many iron foundries existing at the time. With the exception of buildings by James Bogardus, however, the work of his competitors is largely anonymous. But thanks to Badger's enterprise in employing a firm the caliber of Sarony and Major to prepare his catalog, we have preserved for us an extraordinarily complete and delightfully graphic record of the cast iron industry of the time. By means of *Illustrations of Iron Architecture,* we can attribute with certainty many fronts still standing to Badger's concern, and also have knowledge of some that have disappeared. The detail plates give us a close-up view of Badger's shutters, featured in most of the illustrations, and of many other aspects of the finished cast iron front.

While Daniel Badger did everything in his power to exploit his material and made claims for it which the experience of the great city fires of Boston and Chicago would show were extravagant, cast iron still had undoubted advantages for commercial use: first, it was cast and therefore could be mass produced; second, it was strong and light in comparison with masonry. However, because it lacked high tensile strength, cast iron was not capable of long spans save by the agency of trusses and tie rods. As a result, the cast iron front (and this was the characteristic product of the industry), even in its final development in the seventies, was the expression of repetition of the singe window unit, the window being, of course, the wood or iron two over two sash window, or variants thereof. Even though cast iron buildings, or, more generally, buildings having cast iron fronts, might vary in size from one with a twenty-five foot front to one running the whole length of a city block, the scale, based on a standardized unit, remained unaltered, and from the typical twenty-five foot front, divided into three or four equal bays, we can infer what any cast iron front would be like.

This sheer repetition of a standardized unit could sometimes be highly effective, a good example being the Haughwout Building of 1857 (Plate 3, below), still standing at the northeast corner of Broadway and Broome Streets, and now designated a New York City landmark. This building, it should be noted, was the work of an architect, J. P. Gaynor, and it would be valuable to know whether in designing this building he availed himself of "stock" units and combined them as he saw fit; or whether certain portions, at least, were specially cast for such an important commission, and if so, whether the architect's designs, now in the form of molds, were available for reuse. These are only some of the many questions about which we lack information. On the other hand, we do know that many fronts were put up without benefit of an architect, and that stock designs were repeated ov'r and over.

Although iron floor systems with iron supports were developed at the same time as the iron front, their use seems to have been very infrequent. In the smaller buildings floors generally bear on party walls of masonry. In larger buildings something approximating mill construction, with a complete system of internal supports of wood, or iron columns with wood plank floors and beams, would appear to have become almost the rule. Thus the cast iron front was in fact, except for floor heights, practically independent of the construction which stood behind it.

Until the introduction of the elevator the number of floors in commercial buildings seems to have been limited to five. In order to admit more light, the lower stories were higher than the upper ones, which were progressively reduced in height. With window widths, except on the ground floor, not exceeding four feet, and with the width of the individual bay remaining somewhere between six and eight, the cast iron front, in its independence of the construction behind it and its uniformity of scale, is in effect the precursor of the curtain or screen wall of today. Unlike the steel framed structures of the eighties, in which an attempt was made to bring architectonic and structural logic into harmony, the cast iron front largely ignored the structure behind it. Nevertheless, its uniformity and scale, and the deep relief of its columns, arches, or architraves, produced a street architecture rich in the play of light and shadow, especially when seen in sharp foreshortening on a narrow city street.

Every age produces buildings which may be said to belong to no style of art, which exist, so to speak, below the level of architecture; and every age seems at times to build better than it realizes. Daniel Badger's ideas were quite simple. With his team of anonymous architectural designers, modelers, and molders, he sought to reproduce at a lower cost in iron whatever could be produced in stone. If stone had hand cut ornament, iron could achieve the same result by mass production. Iron was strong and apparently fireproof. Its forms could be attenuated without risk of structural failure, and as a result, without sacrificing architectural elegance, more light could be admitted than in more conventional buildings. If the iron front grew drab from accumulations of soot and dirt, all it needed was a coat of paint to freshen it up and make it look like new.

This approach was successful indeed. The iron fronts which had originally been modeled on forms derived from stone were soon the subject of direct imitation in stone itself. Today many iron fronts are hard to distinguish from their attenuated stone competitors. Were it not that the stone ornament has eroded with age and the applied iron ornament has simply fallen off, it would be more difficult still.

Iron and steel construction is of outstanding importance in the history of nineteenth century architecture. While attention has chiefly been drawn to the skyscraper, the development of iron technology which preceded the introduction of steel is of significance even to this day. Daniel Badger's self-assured exploitation of iron as a building material may strike us as being in a way crude, naive, and insensitive, all of which it unquestionably was. Yet in another way—and herein lies the real significance of what he did—Badger was groping toward new ideas of great importance for the future. In building the way he did he built better, perhaps, than he realized.

*Ardsley-on-Hudson, New York*                                                                        WALTER KNIGHT STURGES
*August, 1967*

# BIBLIOGRAPHY

## *Articles*

"America's Cast Iron Age." *Architectural Forum,* Vol. 120 (April, 1964), pp. 110–[113].

Bannister, Turpin C., "Bogardus Revisited." *Journal of the Society of Architectural Historians,* Vol. 15 (December, 1956), pp. 12–22; Vol. 16 (March, 1957), pp. 11–19.

————, "The First Iron-framed Buildings." *Architectural Review,* Vol. 107 (April, 1950), pp. 231–246.

Burnham, Alan, "Last Look at a Structural Landmark" [John Wanamaker Store, N. Y. C.]. *Architectural Record,* Vol. 120 (September, 1956), pp. 273–279.

Fryer, William J. Jr., "Iron Store Fronts." *The Architecture Review and American Builder's Journal,* Vol. 1, (April, 1869), pp. 620-622.

Hitchcock, Henry-Russell, "Early Cast Iron Facades." *Architectural Review,* Vol. 109 (February, 1951), pp. 113–116.

Huxtable, Ada Louise, "Harper and Brothers Building—1854, New York, N. Y." *Progressive Architecture,* Vol. 38 (February, 1957), pp. 153–154.

Sturges, Walter Knight, "Cast Iron in New York." *Architectural Review,* Vol. 114 (October, 1953), pp. 232–237.


## *Source Books*

(Badger, Daniel D.), *Illustrations of Iron Architecture Made by the Architectural Iron Works of the City of New York.* New York, 1865.

Fairbairn, William, *On the Application of Cast and Wrought Iron to Building Purposes.* New York, John Wiley, 1854.

(Thomson, John W.), *Cast Iron Buildings: Their Construction and Advantages,* by James Bogardus. . . . New York, 1856.

*General Works of Reference*

Giedion, Sigfried, *Space, Time and Architecture: The Growth of a New Tradition,* 5th ed. Cambridge, Harvard University Press, 1967.

Gloag, John, and Bridgewater, Derek, *A History of Cast Iron in Architecture.* London, Allen and Unwin, 1948.

Hitchcock, Henry-Russell, *Architecture: Nineteenth and Twentieth Centuries.* Baltimore, Penguin Books, 1958.

Kouwenhoven, John A., *The Columbia Historical Portrait of New York.* Garden City, N.Y., Doubleday, 1953.

# ILLUSTRATIONS
# OF IRON ARCHITECTURE

# ILLUSTRATIONS

OF

# IRON ARCHITECTURE,

MADE BY

## THE ARCHITECTURAL IRON WORKS OF THE CITY OF NEW YORK

———————

NEW YORK:
BAKER & GODWIN, PRINTERS,
PRINTING-HOUSE SQUARE.
1865.

# IRON ARCHITECTURE:

## ITS ORIGIN, ADVANTAGES, AND VARIETY.

## THE ARCHITECTURAL IRON WORKS OF THE CITY OF NEW YORK,

The publishers of this volume, in presenting it to the public, consider it not inappropriate to give a brief account of the introduction of Iron Architecture in this country, setting forth, at the same time, some of the reasons for the superiority of Iron as a building material, and enumerating some of the many forms and uses to which it has been already applied.

It is well known that Iron has been used in England and other European countries for *interior* supports in various kinds of edifices, in the form of columns, beams, etc. ; but its introduction for the *exterior* of buildings is believed to be of purely *American* invention, and of very recent origin.

The first person who practically used Iron as a building material for the exterior was Daniel D. Badger, the President of the Architectural Iron Works.

In the year 1842, Mr. Badger erected, in the city of Boston, the first structure of Iron ever seen in America. The columns and lintels of the first story were of this material, but the prevailing prejudice against this bold innovation was so great that he was not permitted to engage in the work until he had given an ample guaranty that, if it should not prove a success, he would remove it at his own expense.

All the Iron Buildings in this country have been erected since that period, and owe their existence to that humble introduction.

During the following year, A. L. Johnson, of Baltimore, brought to the notice of Mr. Badger his invention of Rolling Iron Shutters. For the purpose of using these shutters, it became necessary to construct the first stories of stores of Iron pillars and hollow posts. At once the superiority of the "Badger Fronts" (as they were then called), in all buildings where large and attractive show windows were desirable, was universally conceded, prejudice began to yield, the manufacture increased, and step by step new and more complete and elaborate designs and improvements came into being, until at last, Iron Architecture became legitimately tested and established.

Mr. Badger transferred the manufacture from Boston to New York, but in a short time it became evident, from the increasing demand for his structures, that greater facilities for their preparation were needed, and the foundation was laid for the present extensive works of this Corporation, situated on Thirteenth and Fourteenth Streets, between Avenues B and C.

Previous to Mr. Badger's introduction of Iron as a building material for exteriors, it is well known that the late Cyrus Alger, Esq., of Boston, had, about the year 1830, made plans and contemplated the erection of an Iron Dwelling House, and that he frequently had stated his conviction of its practicability, and expressed his belief that Iron would in time be adopted as the best material for first-class buildings, on account of its durability and of the beautiful forms of which it was capable, and even upon the consideration of its economy.

It is also known that about the time of Mr. Badger's introduction of this species of building in Boston, one William V. Picket published a volume in London, which he styled "A New System of Architecture, founded on the Forms of Nature, and developing the Properties of Metals."

In the preface to that volume he considered "The Capability of Metallic Bodies for the Realization of Peculiar Beauty," and set forth the value of Iron in Civil as well as Naval Architecture, on account of its "strength, durability, non-combustion, economy of space, facility of construction, and general comfort and convenience, combined with cheapness," and stated his belief that these properties would recommend the application of Iron "in the erection of dwellings and other buildings on land."

The allusion to this work of Mr. Picket is made not for the purpose of elucidating the principles of Architecture laid down by him, for his ideas would be deemed crude at the present time, but simply with the design of showing how recently the subject was regarded as so *novel* as to be claimed as a "New System of Architecture," requiring time for its introduction. Mr. Picket's work was purely theoretical, and we claim, therefore, that for the use of Iron in a practical form the world is largely indebted to Mr. Badger, who may justly be regarded as the inventor and pioneer of Iron Architecture in this country.

That a great change has been wrought in public opinion on this subject since the year 1842 will be evident when the fact is stated that it was with extreme difficulty that owners of property could, at the outset, be induced to employ Iron; the prevalent opinion being that it could not have sufficient strength to support a superstructure unless it was cast solid, and quite as cumbrous as stone, in which case its cost would have been an insuperable objection.

But, by the perseverance of years, this objection and all others were overcome, all prejudices were removed, and to day the practicability of the use of Iron for all kinds of structures is no longer doubted, even by those who were once the most skeptical.

Among those properties of Iron which commend it to more general use as a building material we may mention the following:

## STRENGTH.

The established superiority of Iron in this regard now requires no argument. We may safely affirm that no substance, available for building purposes, has such closeness of texture, or is equally capable of resisting immense pressure.

The great strength of Iron secures another requisite in building, namely:

## LIGHTNESS OF STRUCTURE.

A light and ornamental edifice of Iron may safely be substituted for the cumbrous structures of other substances, and sufficient strength be secured without the exclusion of the light—which is often highly desirable both for mercantile and mechanical purposes.

Combined with this we may mention

## FACILITY OF ERECTION.

Nearly all the work of an Iron structure can be previously prepared and fitted in the foundry and finishing departments, and thence transferred to the place of erection and put together with rapidity and security. In some kinds of structures the facility of erection approaches the incredible.

As has been already mentioned, Iron is capable of all forms of

## ARCHITECTURAL BEAUTY.

It must be evident that whatever architectural forms can be carved or wrought in wood or stone, or other materials, can also be faithfully reproduced in iron. Besides, iron is capable of finer sharpness of outline, and more elaborate ornamentation and finish; and it may be added that it is not so liable to disintegration, by exposure to the elements, as other substances.

To this capability of beauty we may add that of

## ECONOMY OR CHEAPNESS.

The cost of highly-wrought and beautiful forms in stone or marble, executed with the chisel, is often fatal to their use; but they may be executed in Iron at a comparatively small outlay, and thus placed within the reach of those who desire to gratify their own love of art, or cultivate the public taste.

It may also be stated that no other material is so valuable for rebuilding, as Iron always has a market value, and may be recast into new forms, and adapted to new uses. Those who study economy in building should have regard to the permanence of the structure and intrinsic value of the materials, as well as the prime cost of erection.

In an eminent degree Iron possesses the property of

## DURABILITY.

It may be safely affirmed that no material employed for building has such indestructibility as Iron, and none can so successfully resist the wasting influences of the elements. It is also invaluable because of its

## INCOMBUSTIBILITY.

As a resistant of fire, Iron is unequaled. Wherever it is used, the cost of insurance against fire will be materially reduced; and it must be evident that by its use a building may be made absolutely fire-proof. We shall have a better claim to be considered a civilized people when we protect ourselves from the ravages of fire as well as lightning, and erect private and public buildings which are incombustible.

Destructive conflagrations in crowded cities are often arrested by fire-proof buildings, which serve as absolute barriers to the farther progress of the devouring element.

To the catalogue of the excellencies of Iron as a building material may be added its capability of

## RENOVATION.

The durability of an Iron structure is such that if it becomes defaced by exposure or age, it can easily be restored to its pristine beauty by a coating of paint, and, on account of its non-absorbent surface, at less expense than structures of wood or other materials. The color also may from time to time be changed at the will of the owner.

———

THE ILLUSTRATIONS contained in the present volume will show the manifold purposes for which Iron has been applied as a building material, and also exhibit the high degree of architectural beauty which has already been attained.

Special reference to the numerous plates will show that a large number of

## FIRST-CLASS STORES

In the cities of New York, Brooklyn, Philadelphia, Boston, Baltimore, New Orleans, Charleston, Mobile, Memphis, Chicago, and in fact in all large cities and towns, have their fronts built of Iron, ornamented in the most elaborate and varied styles of Architecture—the doors and windows of which are protected by the universally approved

## FIRE AND BURGLAR-PROOF IRON ROLLING SHUTTERS.

These patented Shutters have been extensively used and thoroughly tested for a period of years throughout the country.

The style of Shutters made and introduced by the Architectural Iron Works is considered superior to any other in point of construction as well as price. The gearing is simple, durable, and not liable to derangement. Reference to Plates Nos. 29, 69, and 71, will show the construction and finish in detail.

It may be added that the demand for these Shutters has been so great that this Company is provided with the most complete and elaborate machinery for their speedy manufacture.

Special attention is called to the use of iron for the construction of completely fire-proof buildings to be occupied as

## MANUFACTORIES,

In which strength, solidity, light, and ornament may be combined, and where the necessity of insurance against fire may be obviated. In all large cities such buildings should abound.

On pages 4, 5, and 6, will be found representations of a building of this class, situated on Mott Street, between Broome and Spring Streets, New York. This building, which was erected for I. M. Singer & Co., is six stories in height, with basement and cellar, and is throughout completely fire-proof.

An inspection of this building is needful to give an adequate idea of the solidity of its structure, and of its peculiar fitness for a manufactory.

The protection of life and property afforded by buildings of this class is alone a sufficient warrant for the slightly increased cost of their construction.

Iron has also been successfully used in the erection of

## GRAIN WAREHOUSES.

The amount of the annual losses of grain and warehouse property by fire almost transcends belief. The ordinary storehouses are built wholly or in part of wood, and from certain well-known causes are peculiarly combustible, and liable to rapid decay. Such liabilities are entirely removed by the use of Iron.

The first Iron building of this character was erected in Brooklyn, for " The United States Warehousing Company."

The diagrams on pages 60, 61, and 62 will show that this structure was intended to be used for elevating, transferring, and storing grain, and protecting it against fire.

In the building referred to, the Bins, which are cylindrical, are made like boilers, of riveted plate iron; indeed, the entire structure is absolutely fire-proof and indestructible. Besides these advantages, the grain is secured from the ravages of animals and insects, and also protected from heating by arrangements made for its drying and ventilation. This single feature of ventilation is invaluable, as it will save thousands of bushels of grain which for the want of proper cooling would have to be sent to the malter's at ruinous prices. The construction of Grain Bins of this character is secured to this Corporation by letters patent.

By reference to page 12, it will be seen that Iron has been successfully used for the building of

## ARSENALS

For storing Arms and Ammunition. The first building of this kind was erected in West Troy, in 1858. Safety and durability were the considerations which led to its construction, and it may safely be added that it is admirably adapted for its purposes, and is considered as having secured the objects contemplated in its erection.

## IRON FERRY HOUSES

Have also been constructed by the Architectural Iron Works for the Union Ferry Company of New York and Brooklyn. These structures are an ornament to the city, and supply the place of the unsightly wooden buildings formerly occupying their position, which were liable to rapid decay and destruction by fire.

## IRON OIL TANKS,

Of large capacity, capable of holding hundreds of barrels each, have been constructed for the Phenix Warehousing Company of New York, for the storage of Petroleum.

These tanks, of which many are placed under one roof, are designed to prevent loss by leakage and evaporation, and to protect the oil from the perils of fire. It is believed that they are admirably fitted for all the purposes for which they were constructed.

We would call attention to the combination of

## CAST AND WROUGHT-IRON BEAMS,

As shown in PLATE LXIII. The great strength and elasticity of these Beams consist in their peculiar shape, the necessary quantity of iron being in the proper place, and the Wrought-Iron Tension-rod in the best position to sustain heavy burdens.

This is believed to be the only Beam which prevents all oscillation or trembling of the floors in buildings used for heavy or rapid-running machinery.

A recent and most successful use of Iron has been made by the "Iron Blind Company," in the construction of

## VENETIAN BLINDS,

Both for the outside and inside of windows.

These Blinds present a much lighter appearance to the eye than those made of wood, over which they possess several important advantages. When opened, they admit more light than wooden Venetian Blinds; when closed, they exclude the light more perfectly; they also occupy less space, are more durable, and are proof against fire. They are not liable to warpage or shrinkage, and hence will remain for a long time in working order. They are highly approved by all who have seen or used them. This Corporation has the sole right to manufacture them.

In this connection may be mentioned the use of Iron in the construction of

## WINDOW SASHES,

Which, while scarcely heavier than wood (being hollow), possess the superior advantages of beauty, durability, and incombustibility. These Sashes are especially adapted to warm climates.

It would occupy a large space to enumerate all the uses to which Iron has been applied by the Architectural Iron Works, but the following may be added to those already mentioned, namely: Bridges, Roofs, Domes, Railings, Verandahs, Balustrades, Cornices, Stairways, Columns, Capitals and Arches, Window Lintels and Sills, Consoles, Brackets and Rosettes, Urns, Door and Window Guards, Lamp, Awning and Horse Posts, Patent Lights and Iron Sidewalks.

Reference to the Table of Contents will show numerous uses besides those already enumerated.

———————◆———————

This volume is published at a great cost, for the twofold purpose of supplying Architects and others with plans and details for the construction of the various parts and connections of Architectural Iron Structures, and as an advertising medium for the Architectural Iron Works; and it is designed to be presented to those who may be profited by its study, and aid in the object of extending the business of the publishers, and improving the public taste.

# ILLUSTRATIONS.

XXXI. Elevations and Plans of First-story Fronts.

Nos. 50 and 51.   Arranged for two stores, with Rolling Iron Shutters.
No. 52.                   "        one    "     "     "     "     "
No. 53.   Arranged for two stores, without Rolling Iron Shutters.
No. 54.          "        one store, with          "     "     "   Iron Sash, &c.
No. 55.          "          "          "   passage to lofts, all enclosed with Rolling Iron Shutters.

These designs are on the scale of one-eighth of an inch to the foot, and are intended for 25-feet lots.   They can be adapted to any sized lot.

XXXII. No. 98.   Cornice, with Buttress and Corbel.
No. 50.   Top Cornice.   (See PLATE IX, Gilsey Building.)
No. 112. Arch and Key, with Panel, &c.   (See PLATE LIV.)
No. 115. Tracery Arch Ornament.   (See PLATE LVIII.)

XXXIII. Elevation of First-story Front, with Basement Posts and Piers, with Section and Plan of Side-walk, showing Beams, Girders, &c.

XXXIV. Elevation and Section of Two-story Fronts.

XXXV. No. 33.   Elevation of two stories of store 98 Broadway, New York, showing Rolling Iron Shutters in first story, and Panels on face of doors.
No. 34.   Elevation of first story Nos. 117 and 119 Nassau Street, New York, showing Entrance to Lofts.

XXXVI. Designs for Four-story Fronts.
No. 42, with Pilasters and Arches.
No. 48, with Columns and Antaes and Arches.
No. 46, with Pilasters and Antaes with Arches.
All arranged with or without Rolling Iron Shutters.

XXXVII. No. 17.   Elevation of two stories, with Stone or Brick above; Posts, &c., arranged for Inside Folding Shutters.
No 18.   Elevation of Two-story Front, erected in Augusta, Ga., showing Balustrade in second story.

XXXVIII. No. 9.   Elevation of Five-story Front, erected for R. A. & G. H. Witthaus, No. 38 Barclay Street, New York.
No. 10.   Elevation of Five-story Front, in Gothic Architecture.

XXXIX. No. 106. Elevation and section of Arch, with Key elevation and section of Pier under Arch.
[See PLATE XC.
No. 110. Elevation of section of Arch with Key.   (See PLATE XL, No. 67.)

XL. No. 67.   Elevation of First-story Front, erected in Congress Street, Boston.
No. 68.   Elevation of First-story Front, No. 267 Bowery, New York.

XLI. Window Lintels.   For sizes, &c., see Catalogue of Details.

XLII.      "      "        "        "        "          "

# CATALOGUE OF DETAILS.

| Cornice. | Enriched Modillions. | Ornamented | Plate. XXIII. | No. 95 |
|---|---|---|---|---|
| " | 130 B'way. Enriched Modill. | Dentils. | " | 96 |
| " | B'way & Grand. " | Dentils | XXXII. | 98 |
| " | Gilsey. " | Trusses | " | 52 |
| " | | Plain. Dentils | XLIV. | 88 |

Many other designs can be made with or without enrichments.

**Console or Truss.**

| | Height, including caps, but exclusive of foot leaf. | Project'n. | Width. | Plate. | No. |
|---|---|---|---|---|---|
| Gilsey | 1.4 | .8 | .5½ | X. | 85 |
| 501 Broadway. | 2.6½ | 1.5 | .6 | XVI. | 86 |
| 130 " | 3.8 | 1.4 | 1.0 | XXIII. | 96 |
| | .7 | .5 | .6 | XLII. | 71 |
| | 1.10 | .9 | .10 | XLVII. | 38 |
| | 1.7 | .10½ | .8 | " | 32 |
| | 1.7 | .10½ | .10 | " | 32 |
| | 1.4½ | 1.2 | 1.0 | " | 47 |
| | 1.10 | 1.6 | 1.5 | similar to 47 | |
| | 1.11 | .10½ | .8 | XLVII. | 34 |
| Tiffany | 1.5 | 3.5 | .9 | " | 67 |
| Thomas | 2.6½ | 1.11 | .6 | " | 53 |
| Tiffany | 1.8½ | 1.7 | .8 | " | 46 |
| | 2.3 | 1.2 | 1.2 | " | 8 |
| | 2.3 | 1.2 | 1.0 | " | 8 |
| Similar to No. 8, but no carved mold'g on face. | 2.3 | .9 | .10 | | K |
| | 2.3 | .9 | 1.0 | | K |
| | 2.3 | .9 | 1.2 | | K |
| Brandreth | 2.9 | 1.5 | 1.4 | XLVII. | 35 |
| | 2.1 | .9 | .6 | XLVIII. | 275 |
| | 2.1 | .9 | .8 | " | 275 |
| | 2.1 | .9 | .10 | " | 275 |
| | 2.1 | .9 | 1.0 | " | 275 |
| | 2.1 | .9 | 1.2 | " | 275 |
| | 2.6 | .11 | .6 | " | 277 |
| | 2.6 | .11 | .8 | " | 277 |
| | 1.8½ | .7 | .7 | " | 279 |
| | 1.8 | .5 | .7 | " | 279 |
| Similar to 276 | 2.4 | 1.2 | .6 | | G |
| " " | 2.0 | 1.0 | .8 | | E |
| | 3.0 | 3.0 | .. | XLVIII. | 6 |
| | 1.4 | .10 | .8 | LXXII. | 33 |
| | .. | .. | .. | " | 36 |
| | 2.8½ | .8 | .8 | " | 37 |
| | 1.3 | .7 | .7 | " | 39 |
| Similar to 39 | 1.4 | .6 | .7½ | .. | Q |
| | 1.4 | .10 | | .. | A |
| | 1.3 | .8 | .. | .. | B |
| | 1.8½ | .11½ | .8 | .. | F |
| | 1.8 | .6 | .5 | .. | D |
| | 2.0 | .9 | .9 | .. | H |
| Palmer | 3.2 | 1.7 | .8 | .. | I |
| " | 3.2 | 1.7 | .10 | .. | I |
| | 2.2 | 1.0 | 1.0 | .. | J |
| | 2.2 | 1.0 | 1.2 | .. | J |
| | 2.2 | 1.0 | 1.4 | .. | J |
| | 1.6 | .6 | .5 | .. | L |

| | Height, including caps, but exclusive of foot leaf. | Projection. | Width. | Pl t . | No. |
|---|---|---|---|---|---|
| Console or Truss | 1.6 | .6 | .6 | .. | L |
| | 1.8 | .4 | .4 | .. | M |
| Face on front. | 2.0 | .5 | .6 | .. | N |
| | 2.0 | .5 | .7 | .. | O |
| | 2.0 | .5 | .7 | .. | P |
| | 1.7 | 1.0 | .8 | .. | R |
| | 1.6 | 1.0 | .8 | .. | S |

Many other designs and sizes.

| | Plate | No. |
|---|---|---|
| Dentils | V. | 119 |
| " | VIII. | 117 |
| " | " | 118 |
| " | X. | 97 |
| " | XIII. | 120 |
| " | XVIII. | 100 |
| " | XX. | 87 |
| " | " | 93 |
| " | XXIII. | 96 |
| " | XXXII. | 98 |

Also, other designs.

| | Plate | No. |
|---|---|---|
| Gates | XCVIII. | 245 |
| " | XCIX. | 257 |

Also, many other designs.

| | Plate | No. |
|---|---|---|
| Girder, Arch and Tension Rod | LXIII. | 271 |
| " " " | " | 272 |
| " " " | " | 273 |
| " " " | " | 274 |

| | | Height. | Width. | Plate | No. |
|---|---|---|---|---|---|
| Guard, Door. | Lattice | 4.4½ | .10 | LXXVII. | 131 |
| " | " | 4.3½ | 1.1 | " | 131 |
| " | " | 4.1½ | .10 | " | 131 |
| " | " | 3.9 | 1.1 | " | 132 |
| " | " | 3.9 | .11½ | " | 132 |
| " | " | 3.9 | .10 | " | 132 |
| " | " | 3.5½ | 1.1 | " | 132 |
| " | " | 3.5½ | .11½ | " | 132 |
| " | " | 3.5½ | .10 | " | 132 |
| " | " | 3.1½ | .10 | " | 132 |
| " | " | 4.3½ | 1.1 | " | 130 |
| " | " | 3.9 | 1.1 | " | 141 |
| " | " | 3.9 | .11½ | " | 129 |
| " | " | 3.9 | .11½ | " | 127 |
| " | " | 4.3½ | 1.1 | " | 137 |
| " | " | 3.9 | 1.1 | " | 137 |
| " | " | | | XCI. | 48 |
| " | " | | | " | 205 |
| " | " | | | " | 206 |
| " | " any size, with or without Border. | | | " | 203 |
| " | " | | | " | 204 |
| " | " | | | " | 207 |

All with Square or Circular Tops or Bottoms.
Also, other designs.

# CATALOGUE OF THE PRINCIPAL WORKS

### ERECTED BY THE

# ARCHITECTURAL IRON WORKS.

| LOCATION. | PROPRIETOR. | ARCHITECT. | DESCRIPTION. |
|---|---|---|---|
| ALBANY, N. Y., Greene St... | Albert Blair.......... | ................. | 3 Store Fronts. |
| Do    do   N. Pearl St.. | James Kidd............ | ................. | 2    " |
| Do    do     do | do   .......... | —— Smith...... | 42 feet 4-story Front, similar to Plate XXI. |
| Do    do   .......... | A. Koonz............. | | 1 Store Front. |
| Do    do   .......... | Woollett & Ogden...... | | 1    " |
| ALEXANDRIA, La.......... | | | Court House Portico, and Main Course. |
| ALEXANDRIA, Egypt........ | R. H. Allen & Co...... | | Iron Storehouse. |
| ALLEGHANY CITY, Penn.... | Gordon & Rafferty.... | | 1 Store Front. |
| ATLANTA, Ga............. | I. Boutell............ | | 80 feet Store Front. |
| Do   ............. | Beach & Root......... | | 49   "     " |
| AUBURN, N. Y............ | J. W. Haight......... | | 1 Store Front. |
| Do   ........... | F. L. Griswold & Co... | | 1    " |
| AUGUSTA, Ga............. | T. S. Metcalf........ | | 15 Store Fronts. |
| Do   ........... | City Bank ........... | | Exterior Iron Work. |
| Do   ........... | Lambeck & Cooper..... | | 2-story Front.   Plate XXXVII., No. 18. |
| BALTIMORE, Md........... | Baltimore Sun........ | R. G. Hatfield...... | 7 Store Fronts. |
| Do   ........... | Canfield, Brothers & Co. | ................. | 1    " |
| Do   ........... | J. King............. | J. Dixon .......... | 1    " |
| BATH, N. Y. ............ | H. W. Perrine ....... | M. Austin......... | 1    " |
| Do   ........... | A. S. Howell.......... | G. E. Bartlett...... | 1    " |
| BOSTON, Blackstone Street... | Ritchie & Wentworth... | ................. | 50-feet Store Fronts. |
| Do    do   do  .. | Mr. Richardson........ | S. P. Fuller........ | 87   "     sim. to Pl. XXVIII., No. 62. |
| Do  Congress   do  .. | L. Ware ............. | G. J. F. Bryant.... | 109   "     "     " XL., No. 67. |
| Do  Court   do  .. | J. C. Gray .......... | C. K. Kirby ...... | 50-feet 5-story Front.   See Plate LVIII. |
| Do  Federal   do  .. | Mr. Kramer.......... | G. J. F. Bryant.... | 1 Store Front.   Plate XL., No. 67. |
| Do  Theatre Alley...... | C. Merriam & Sons..... | S. P. Fuller....... | 5 Store Fronts.   Plate XXVIII., No. 62. |
| Do  Washington St..... | H. H. Hunnewell...... | George Snell...... | 5-story Front.   Plate LIX. |
| BRANT, Canada West....... | Hegeman & Co......... | | 2 Store Fronts. |
| BRIDGEPORT, Conn........ | S. Sterling.......... | | 1 Store Front. |
| Do    do   ........ | R. E. Stanton........ | Lambert & Bunnell. | 1    " |
| Do    do   ........ | City Bank ........... | do | 40-feet 3-story Front. |
| Do    do   ........ | Benham, Brothers...... | do | 2 Store Fronts. |
| BROOKLYN, N. Y ......... | Phenix Warehouse Co.. | ................. | 42 Oil Tanks, 300 bbls. each. |
| Do   do  Atlantic St. | J. Smith............. | King & Kellum .... | 1 Store Front. |
| Do   do  Atlantic D'k. | U. States Warehouse Co. | G. H. Johnson & Co. | Grain Warehouse, Plates LX., LXI., LXII., 107 × 125 feet.   5 stories.   Fire-proof. |

| LOCATION. | PROPRIETOR. | ARCHITECT. | DESCRIPTION. |
|---|---|---|---|
| BROOKLYN, N. Y., Fulton St. | Mr. Williams | | 2 Store Fronts. |
| Do do do | E. Lewis | | 1 " |
| Do do do | L. J. Horton | | 2 " |
| Do do do | J. O. Whitehouse | | 1 " |
| Do do do | Smith & Jewell | G. H. Johnson | Iron Works, Fulton Mills. |
| Do do do | City of Brooklyn | | 3 Engine House Fronts. |
| Do do do | V. G. Hall | | 1 Store Front. |
| Do do do | W. H. Cary | King & Kellum | 3 Store Fronts. |
| Do do do | Mr. Newman | Mr. Roberts | 1 Store Front. |
| Do do do | J. Burroughs | King & Kellum | 1 " |
| Do do do | J. Halsey | G. H. Johnson | Halsey Building.   Plate LII. |
| Do do do | C. V. B. Ostrander | | 1 Store Front. |
| Do do do | Smith & Jewell | G. H. Johnson & Co. | 75 feet " |
| Do do do | Kings County | King & Tecknitz | Iron Work, Kings Co. Court Ho. : Beams, Roof, Dome, Stairs, Sashes, Shutters, &c. |
| Do Hamilton Av | Smith & Jewell | G. H. Johnson | Iron Work, Atlantic Flour Mill. |
| Do Navy Yard | U. S. Government | | "     Government Stores. |
| Do Pierrepont Place. | A. A. Low | F. A. Petersen | 47 feet Greenhouse Front. |
| Do do | do | do | 1 Verandah. |
| Do do | A. A. White | do | 1 " |
| BUFFALO, N. Y | George Coit, Jr. | | 2 Store Fronts. |
| Do | Mr. Buckley | | 1 " |
| Do | Brown, Brothers | | 173 feet Front, Brown's Building. |
| CHARLESTON, S. C. | Dwing, Thayer & Co | | 1 Store Front. |
| Do | O. M. Cohen | | 1 " |
| Do | F. O. Fanning & Co | | 1 " |
| Do | R. Boyce | | 2 " |
| Do | H. W. Conner | | 1 " |
| Do | Hariel, Hare & Co | | 3 " |
| Do | T. A. P. Horton | | 1 " |
| Do | A. Elfe | | 1 " |
| Do | Bancroft, Betts & Marshall | | 2 " |
| Do | L. M. Hatch | | 1 " |
| Do | W. J. Walker & Brother | | 2 " |
| Do | J. E. Spear | | 1 " |
| Do | P. O'Donnell | | 1 " |
| Do | U. S. Government | | Columns Custom House. |
| CHICAGO, Ill | C. R. Starkweather | | 2 Store Fronts. |
| Do | A. Robbins | J. M. Van Osdell | 231 feet 5-story Front.   Plate LIV. |
| Do | J. Link | do | 150 feet 5- "     " VII. |
| Do | Lloyd and Sons | do | 161 feet 5- "     " XIX. |
| Do | F. Tuttle and others | do | 158 feet 5- "     " LXX. |
| Do | Price, Church & Co | | 150 feet 5- " |
| CUBA | F. P. Dias | | Market, &c. |
| DETROIT, Mich | F. & C. H. Buhl | | 4 Store Fronts. |
| DUNKIRK, N. Y | E. Risley & Co | | 5 " |
| FORT LAFAYETTE | U. S. Government | | Iron Work. |
| GEORGETOWN, S. C. | S. W. Ronguie | | 31 feet Store Front. |
| GLENHAM, N. Y | Glenham Company | | 1 Store Front. |
| GRAND RAPIDS, Mich. | J. W. Pierce | | 1 " |
| Do do | W. P. Collins | | 2 " |
| HALIFAX, N. S | W. G. Combes | | 4-story Front.  Plate LXXIX. |
| Do | Mr. Bennett | C. P. Thomas | 1 Store Front.   " LXXVI. |
| Do | Duffries & Co | do | 2 "     " LXXVIII. |
| Do | Mr. Coleman | do | 2 "     " LXXV. |
| Do | Mr. Billings, Jr. | do | 1 "     " LXVIII. |

| LOCATION. | PROPRIETOR. | ARCHITECT. | DESCRIPTION. |
|---|---|---|---|
| HALIFAX, N. S. | Mr. Chipman | C. P. Thomas | 2 Store Fronts, Plate LXXIV. |
| Do | Mr. Skerry | do | 1 Store Front, " LXXIV. |
| Do | Mr. Mignowitz | do | 1 " " LXXIV. |
| Do | Mr. Billings | do | 2 Store Fronts. " LXXIV. |
| Do | C. C. Tropolet | | 3 " |
| Do | Mr. Roman | C. P. Thomas | 2 " |
| Do | Mr. Scott | | 2 " |
| HAVANA, Cuba | Pesant, Brothers | | Sugar Sheds, 143 feet long.  Pl. LXXIII. |
| Do | Spanish Navy | | Lumber Sheds, 81 × 120 feet. |
| HIGH BRIDGE | Croton Aqueduct Dept. | | Railing, High Bridge. |
| LANCASTER, Penn | J. N. Lane & Nephews | | 1 Store Front. |
| LOCKPORT, N. Y | N. J. Dunlap | | 1 " |
| LYNCHBURG, Va | J. T. Davis | | 1 " |
| Do | W. S. Ellison | | 1 " |
| MARTINSVILLE, La. | Tertron, Bronsard & Co. | | 1 " |
| MATANZAS, Cuba | C. A. Caruano | | Columns, &c., Public Hall. |
| MEMPHIS, Tenn | Mosely & Hunt | | 4-story Front, similar to Pl. XV., No. 7. |
| Do | W. B. Greenlow | | 2 Store Fronts.  Plate XXXIII. |
| Do | Cooke & Co | Fletcher & Wintter. | 4-story " " XLVI. |
| MILWAUKEE, Wis | H. J. Nazro & Co | | 2 Store Fronts. |
| Do | Mack, Ottinger & Co | Otto Schwartz | 1 Store Front. |
| Do | Mahler & Wendt | | 2 " |
| Do | J. B. Martin | G. H. Johnson | 160 feet 4-story Front.  Plate XLVI. |
| MOBILE, Ala | J. Emanuel | | 2 Store Fronts. |
| Do | Daniels, Elgin & Co | J. H. Giles | 45 feet 4-story Front. |
| Do | do | do | 103 feet Store Front. |
| NEWARK, N. J | Mr. Dennis | Mr. Hall | 1 Store Front. |
| Do | J. McGregor | | 40 feet 4-story Front.  Plate XXVI. |
| Do | J. W. Corey | | 1 Store Front. |
| NEW HAVEN, Conn | H. N. Whittlesey | | 2 Store Fronts. |
| Do | T. Bennett | | 2 " |
| Do | Perkins, Treat & Chatfield | | 2 " |
| Do | Young Men's Inst. | | 2 " |
| Do | A. Parker | | 3 " |
| NEW LONDON, Conn | S. & G. Rogers | | 2 " |
| NEW ORLEANS, La | Paul Tulane | | 47 feet 5-story Front.  Plate VII. |
| Do | J. B. Lee | | 62 feet Store Front. |
| **NEW YORK:** | | | |
| Barclay St., No. 34 | R. A. & G. Witthaus | S. A. Warner | 5-story Front.  Plate XXXVIII., No. 9. |
| Do  36, 38 | | T. R. Jackson | 2 Store Fronts. |
| Do  50 | Mr. Gibson | J. C. Wells | 1 " |
| Do  52, 54 | Wolfe & Mickle | | 2 " |
| Do  58 | T. E. Gilbert | W. H. Hume | 1 " |
| Beekman St.,  23, 25 | | | 2 " |
| Do  27 | | | 1 " |
| Do  29 | Remsen & Ely | G. W. Noble | 1 " |
| Do  55, 57 | | J. B. Snook | 2 " and Rears. |
| Do  79, 81 | | Do | 2 " |
| Do  83 | | Do | 5-story Front. Plate XC., No. 14. |
| Bleecker St., cor. Mercer. | Mr. Bosch | | 1 Store Front. |
| Do  do | Judge Jackson | Kellum & Son | 35 feet Store Front. |
| Do  do | A. T. Stewart | J. B. Snook | 82 " " |
| Bowery, No. 13 | C. S. Hines | | 1 Store Front and Basement. |
| Do  70, 72 | Wm. B. Astor | | 2 " " |
| Do  96 | A. L. Ely | R. G. Hatfield | 1 " |
| Do  98 | | Thomas & Son | 1 " |

| LOCATION. | PROPRIETOR. | ARCHITECT. | DESCRIPTION. |
|---|---|---|---|
| **NEW YORK:** | | | |
| Bowery, No. 110 | A. L. Ely | R. G. Hatfield | 1 Store Front and Basement. |
| Do 163 | McGraw and Taylor | | 1 " |
| Do 267 | J. B. Simpson | J. B. Snook | 1 " Plate XL., No. 68. |
| Do cor. Canal | Lorillards | Thomas & Son | 132 feet Store Front. |
| Do " Delancey | J. B. Simpson | A. Winham | 50 " |
| Do " Houston | Lorillards | | 4 Store Fronts. |
| Do " Bond | | J. B. Snook | 40 feet " |
| **Bridge St.** | R. Blanco | | 91 " " |
| **Broadway,** Nos. 39 to 49 | McCurdy, Aldrich & Spencer | | 5 Store Fronts. |
| Do 53 | P. & R. Goelet | | 5-story Front and Basem't. Pl. XV., No. 8. |
| Do 61 | | S. A. Warner | 82 feet Store Front. |
| Do 63, 65 | L. S. Suarez | | 2 Store Fronts. |
| Do 70 | James Harriot & Co. | | 1 Store Front. |
| Do 72 | Mr. Cruger | Thomas & Son | 1 " |
| Do 84 | Mr. St. John | R. G. Hatfield | 31 feet Front, 2 story. Pl. XXXV., No.33. |
| Do 102 | Continental Ins. Co. | G. Thomas | 2 Store Fronts. |
| Do 112, 114 | Bowen & McNamee | | Patent Shutters for 2 Fronts. |
| Do 130 | W. H. Hume | W. H. Hume | 5-story Front. Plate XXII. |
| Do 139 | | J. F. Duckworth | 1 Store Front. |
| Do 141 | G. & W. Young | | 1 " |
| Do 151 | | J. F. Duckworth | 1 " |
| Do 156, 158 | D. H. Haight | J. B. Snook | 2 Store Fronts. |
| Do 161, 163, 165 | F. Marquand | | 3 " |
| Do 177 | W. H. Smith | | 1 Store Front. |
| Do 179 | Thomas Hunt | King & Kellum | 1 " |
| Do 187 | Noel J. Becar | | 1 " |
| Do 194 | L. M. Wiley | | 1 " |
| Do 199 | —— Young | | 1 " |
| Do 202 | A. Cleveland | Thomas & Son | 1 " |
| Do 203 | J. Q. Jones | | 1 " |
| Do 204 | Appletons | Thomas & Son | 1 " |
| Do 205 | W. H. Smith | | 1 " and Basement. |
| Do 235 | Tracy, Irwin & Co. | Thomas & Son | 1 " |
| Do 241, 243 | Solomon & Hart | F. A. Petersen | 2 Store Fronts. |
| Do 245 | D. O'Connor | | 1 Store Front. |
| Do 252 | | | 5-story Front, similar to Plate XV., No. 7. |
| Do 257 | Mr. Field | | 1 Store Front. |
| Do 287 | S. Storms | | 1 " |
| Do 292, 296 | J. De Forest | | 2 Store Fronts. |
| Do 294 | B. Pike, Jr. | | 1 Store Front. |
| Do 300 | W. B. Astor | G. Thomas | 1 " |
| Do 306 | Mr. Barclay | | 1 " |
| Do 332 | John Dolan | J. B. Snook | 5-story Front, similar to Plate XC. |
| Do 341 | Paton & Co. | J. H. Giles | 1 Store Front. |
| Do 343 | Geo. Ponsot | | 1 " |
| Do 349 | J. & I Cox | | 1 " |
| Do 355 | Adriance & Strang | | 1 " |
| Do 360 | P. Lorillard | | 1 " |
| Do 361, 363 | H. Wood | J. Rogers | 2 Store Fronts. [No 66. |
| Do 369 | Solomon & Hart | F. A. Petersen | 1 Store Front and Basement. Pl. XXVIII., |
| Do 371, 375 | Dr. Moffat | | 2 " " and Rear. |
| Do 372 | H. D. Aldrich | S. A. Warner | 1 " |
| Do 373 | L. Spencer | J. B. Snook | 5-story Front and Basement. Plate XIX. |
| Do 377, 379 | Mr. Lawrence | | 2 Store Fronts. |

| LOCATION. | | PROPRIETOR. | ARCHITECT. | DESCRIPTION. |
|---|---|---|---|---|
| NEW YORK: | | | | |
| Broadway, No. | 388 | D. Wood | King & Kellum | 1 Store Front, Basement and Rear. |
| Do | 403 | P. & R. Goelet | | 1 " |
| Do | 404, 406, 408 | P. Lorillard | | 3 Store Fronts. |
| Do | 405 | Duncan & Sons | | 1 Store Front. |
| Do | 442 to 454 | G. W. Miller | A. Winnam | 125 feet Store Fronts. City Assembly R'ms. |
| Do | 447 | Mr. Collamore | | 1 Store Front. |
| Do | 449 | Mr. Jackson | | 1 " |
| Do | 452, 454 | Peter Goelet | | 1 " |
| Do | 456 | T. Woodruff | | 1 " |
| Do | 471 | W. Gibson | | 1 " |
| Do | 481 | J. De Wolfe | | 1 " |
| Do | 495 | Grover & Baker | G. H. Johnson | 3-story Front. Plate XI. |
| Do | 501 | O. B. Potter | Thomas & Son | 5-story Front. Sim. to Plate XC. |
| Do | 502, 504 | Dr. H. Bostwick | Kellum & Son | 2-store Fronts. Basement and Rear. |
| Do | 506 | E. Langdon | | 1 Store Front. |
| Do | 508 | | | 1 " |
| Do | 516 | Savings Bank | | Pat. Shutters for Front |
| Do | 535 | S. Brewster | | 1 store Front. |
| Do | 543 | W. B. Astor | | 1 " |
| Do | 547 | | | 1 " |
| Do | 550 | Tiffany & Co | R. G. Hatfield | 1 " and Base't. Plate LXIV., No. 26. |
| Do | 552, 554 | R. French | J. B. Snook | 2 " |
| Do | 555 | A. R. Eno | Thomas & Son | 1 " |
| Do | 577, 579, 581 | Langdon family | | 3 " |
| Do | 585 | | J. Rogers | 1 " |
| Do | 601 | W. & E. Mitchell | | 1 " |
| Do | 604 | | | 1 " |
| Do | 606 | | | 1 " |
| Do | 620 | Henry Dolan | J. B. Snook | 6-story Front. Sim. to Plate VII. |
| Do | 621 | Gerard Stuyvesant | | 1 Store Front. |
| Do | 624, 626 | Uhl & Whitney | J. M. Trimble | Ent. to L. Keene's Thea. Pl. LXIV., No. 24. |
| Do | 627, 629 | S. Brewster | | 2 Store Fronts. |
| Do | 631 to 637 | P. & R. Goelet | | 4 " |
| Do | 645 | Mr. Agate | | 1 Store Front. |
| Do | 653 | | J. F. Duckworth | 1 " |
| Do | 654 | | | 1 " |
| Do | 667 to 677 | J. La Farge | Jas. Renwick | 150 ft " Old La Farge House. |
| Do | " | " | Jas. Renwick | 150 ft " New " " |
| Do | 679 | | | 1 " |
| Do | 701 | Mr. Manice | | 1 " |
| Do | 706 | | | 1 " |
| Do | 711 | Mr. Holmes | | 1 " |
| Do | 747 | J. G. Pearson | | 1 " |
| Do | 752 | N. Y. Dyeing & P. Est. | | 1 " |
| Do | 758 | S. Kohnstamm | Kellum & Son | 1 " |
| Do | 785 | J. Colles | | 1 " |
| Do | 845 | C. V. S. Roosevelt | | 1 " |
| Do | 847 | " | | 1 " |
| Do | cor Exch. Pl. | J. Steward, Jr., & Co | | 1 " |
| Do | " Pine | Continental Ins. Co. | Thomas & Son | 2 Fronts. |
| Do | " Cedar | | J. B. Snook | 2 store Fronts. |
| Do | " Liberty | Mr. Herrick | | 1 Store Front. |
| Do | " Cortlandt | | | 1 " |
| Do | " " | P. Gilsey | J. W. Ritch | 6-story Front. 163 ft. Gilsey Build., Pl. IX. |
| Do | " Dey | W. W. Chester | | 1-store Front. |

| LOCATION. | | PROPRIETOR. | ARCHITECT. | DESCRIPTION. |
|---|---|---|---|---|
| **NEW YORK:** | | | | |
| Broadway, cor. Murray | | Ball, Tompkins & Black | | 1 Store Front. |
| Do | " Warren | S. V. Hoffman | J. B. Snook | Pat. Shutters, 150 feet Front. |
| Do | " Chamber and Reed. | A. T. Stewart | J. B. Snook | " |
| Do | " Pearl | J. Gemmell | S. A. Warner | 50 feet Store Front. |
| Do | " Anthony | Dr. J. Moffat | | 1 Store Front. |
| Do | " Pearl&Worth | J. R. Whiting | Kellum & Son | 175 feet Store Front and Basement. |
| Do | " Leonard | Appletons | | 2 Store Fronts. |
| Do | " " | | | 1 " |
| Do | " " | W. G. Lane & Co | | 198 feet Store Front. |
| Do | " Franklin | W. Gibson | | 1 Store Front. |
| Do | " " | John Taylor | Thomas & Son | 75 feet Front. Taylor's Saloon. |
| Do | " White | Mr. Clark | | 1 Store Front. |
| Do | " Canal and Lispenard | Dr. Brandreth | Chas. Mettam | 290 feet Store Front and main cornice. Brandreth House. |
| Do | " Grand | B. Wood | R. G. Hatfield | 130 feet Store Front. |
| Do | " Broome | W. Gale & Son | Chas. Mettam | 228 ft. " |
| Do | " " | W. Langdon | J. P. Gaynor | 5-story Front. 162 ft. Haughwout Building Plate III. |
| Do | " Spring | R. H. Haight & Co., | J. B. Snook | 470 ft. " St. Nicholas Hotel. |
| Do | " " | J. J. Astor's heirs | R. G. Hatfield & D. Lemon. | 176 ft. " |
| Do | " Prince | A. T. Stewart | Trench & Snook | 237 ft. " Metropolitan Hotel. |
| Do | " Fourth | Mr. Philbin | | 113 ft. " |
| Do | " Wash'ton Pl. | | | 35 ft. " 2 story. |
| Do | " Ninth street | | | 1 Store Front. |
| Do | " Tenth | J. Beck | R. Henry | 163 ft. " 2 story. |
| Do | " Twelfth | S. Whitney's heirs | | 2 Store Fronts and Basements. |
| Do | " Thirteenth | Mr. Valentine | | 58 ft. " |
| Do | " Twentieth | | | 1 Store Front. |
| Do | " Twenty-first | S. Halsted | King & Kellum | 2 " Pl. XXIV. No. 64. |
| Do | " Madison Sq. | | | 2 " |
| Do | " Twenty-fifth | Mr. Livingston | J. B. Snook | 2 " |
| Do | bet. 25th and 26th | S. V. Hoffman | J. B. Snook | Hoffman Hotel. |
| Do | cor. Twenty-sixth. | Mr. Dodworth | Renwick & Co | 142 ft. Front. |
| Do | | Mr. Bulkley | | 1 Store Front. |
| Do | | W. Snickner | | 1 " |
| Do | | S. Brewster | | 3 " |
| Do | | | | 1 " |
| Burling Slip | | A. A. Low | | 2 Store Fronts. |
| Canal street, 253 | | J. Hesley | Kellum & Son | 1 " and Basement. |
| Do | 262 | P. H. Frost | | 1 " |
| Do | cor Wooster | Rosenblat & Banta | W. T. Beers | 2 " |
| Do | " Thompson | People's Bank | Thomas & Son | 75 ft. " |
| Do | | Mr. White | | 1 Store Front. |
| Do | | N. & J. Brown | | 1 " |
| Do | | | | 1 " |
| Cedar | | J. D .Phillips | | 2 story Front. |
| Central Park | | C. P. Commissioners | Olmstead & Vaux | Bridge Railing. Plate XCV., No. 229. |
| **Chambers Street :** | | | | |
| Do | 43 to 49 | Spencer, Wyeth&Stewart | R. Henry | 4 Store Fronts and Rear. |
| Do | 53 | R. Henry | | 1 " " |
| Do | 76 | | J. M. Rich | 1 " |
| Do | 77 | Peddie & Merriam | | 5-story Front and Basement, similar to Plate XXXVIII, No. 39. |

| LOCATION. | | PROPRIETOR. | ARCHITECT. | DESCRIPTION. |
|---|---|---|---|---|
| **NEW YORK:** | | | | |
| Chambers Street, | 78 | Dr. Alcock | | 1 Store Front. |
| Do | 80, 82 | Mr. Holmes | | 2 " |
| Do | 88 | | | 1 " |
| Do | 113 | | | 1 " and Basement. |
| Do | 120 | W. H. Jones | | 50 ft. Store Front, 5-story, sim. to Pl. XC. |
| Do | 122 | | | 1 Store Front and Basement. |
| Do | 126, 128 | Holmes & Colgate | Thomas & Son | 2 " |
| Do | 152 | | | 1 " and Basement. |
| Do | cor Church | T. Suffern | | 45 ft. " |
| Do | and Warren | H. D. Aldrich | S. A. Warner | 2 Store Fronts. |
| Do | " Reade | A. T. Stewart | J. B. Snook | 218 ft. " and Basement. |
| Do | " " | J. Haggerty | | 2 Store Fronts. |
| Do | " " | W. S. Wetmore | | 240 ft. Front. |
| Do | " " | W. H. Cary | King & Kellum | 100 " 5-story. Pl. VII. |
| Chatham Street, | 19 | J. B. Simpson | J. B. Snook | 1 Store Front. Basement and Rear. |
| Do | 65 | Chatham Bank | J. B. Snook | 1 " |
| Do | 74 | | | 1 " |
| Do | | J. B. & W. Simpson | | 2 " French's Hotel. |
| Church Street, | | S. D. Babcock | T. S. Wall | 1 " |
| " | | Mr. Matthews | | 1 " |
| " | cor Fulton | Mr. Phyfe | J. B. Snook | 108 ft. " |
| Cliff Street, 22 | | Mr. James | | 1 Store Front and Rear. |
| College Place, 7 | | W. H. Grinnell | R. H. Mook | 5-story " and Base't. Pl. XV, No. 7. |
| Cortlandt Street 3 | | C. & U. J. Smith | | 1 Store " |
| Do | 4 | C. Vanderbilt | | 1 " |
| Do | 6, 8 | Brown & Cushing | | 2 " |
| Do | 12, 14, 16 | | Mr. Hurry | 3 " |
| Do | 15, 17 | Mr. Dwire | | 2 " |
| Do | 18 | | | 1 " |
| Do | 19, 21 | | | 2 " |
| Do | 20 | Gilbert, Prentiss & Tuttle | | 1 " |
| Do | 22 | Bennett & Johnson | | 1 " |
| Dey Street, 5, 7 | | Noel J. Becar | | 2 " |
| Do | 13 | Wilson & Co | | 1 " |
| Do | 15 | Mr. Cox | | 1 " |
| Do | 44 | E. H. Main | | 1 " |
| Do | 46 | | | 1 " |
| Duane Street, 42 | | Archl. Iron Works | G. H. Johnson | 5-story " and Base. Sim. to Pl. VII. |
| Do | 68, 75 to 85 | Mr. Palmer | J. B. Snook | 165 ft. " 5-story. Sim. to Pl. XIX. |
| Do | 71, 73 | J. B. Snook | J. B. Snook | 50 ft. " and Basement. |
| Do | 84, | | J. F. Duckworth | 5-story " and Basement. |
| Do | 116 | | | 1 Store Front and Basement. |
| Do | and Reade | | | 2 " |
| Do | cor Church | Dr. Lovejoy | Amzi Hill | 116 ft. " |
| Do | " Reade | East River Bank | | 1 Store Front. |
| Fifth Av. 23d St, B'way & 24th | | A. R. Eno | Thomas & Son | 277 ft. Front. Fifth Avenue Hotel. |
| Fourth Av | | Harlem Railroad Co | | Columns, &c., Engine Depot. |
| Do cor 23d Street | | | Mettam & Burke | 5 Store Fronts. |
| Franklin Street, 73 | | John Mack | Renwick & Co | 1 " Pl. XCII. |
| Do | 91 | " | " | 1 " |
| Do | cor Franklin Pl | | Thomas & Son | 53 ft. " Basement and Rear. |
| Do | " Church | W. Watson | T. S. Wall | 2 Store Fronts. |
| Do | " | " | " | 2 " |
| Front Street | | J. K. & E. B. Place | E. L. Roberts | 1 " |
| " | 166 | | | 1 " |

| LOCATION. | | PROPRIETOR. | ARCHITECT. | | DESCRIPTION. |
|---|---|---|---|---|---|
| NEW YORK: | | | | | |
| Fulton Street, 58, 60 | | | | | 2 Store Fronts. |
| Do | 115 | | | 1 | " |
| Do | 141 | M. Reily | | 1 | " |
| Do | 186 | J. Tucker | | 1 | " |
| Do | 198 | Mr. Phyfe | | 1 | " |
| Do | | Union Ferry Co | Kellum & Son | | Fulton Ferry House. |
| Gouverneur's Lane | | E. Banker & Co | | | 3 Store Fronts. |
| Grand Street, cor. Chrystie | | Lord & Taylor | Thomas & Son | 200 ft. | " |
| Do | " Allen | Mr. Donnelly | T. S. Wall | | 1 Store Front. |
| Greenwich Street, 52 | | | Griffiths | 1 | " |
| Do | 218 | | C. Mettam | 1 | " |
| Do | 282 | | Thomas & Son | 1 | " |
| Do | | A. T. Lagrave | | 1 | " |
| Do | | Mr. Platt | | 1 | " |
| Do | | W. B. Astor | | 1 | " |
| Howard Street, | | | | 1 | " |
| Hudson St. 250 | | Mr. Sloan | | 1 | " |
| Do | 277, 279, 281 | B. Newhouse | | 3 | " |
| Do | 297 | A. M. L. Scott | | 1 | " |
| Do | cor Broome | J. S. Hasbrook | | 1 | " |
| Do | " Jay | Am. Express Co | J. W. Ritch | 204 ft. | " |
| Irving Pl., | " 15th | Manhattan Gas Co | W. W. Gardiner | | 1 Office Front. |
| John St., | 19 | F. W. Lasak | | | 1 Store Front. |
| Do | 20 | Mr. Young | | 1 | " |
| Do | 22 | | | 1 | " |
| Do | 75 | J. K. Herrick | F. Diaper | 1 | " |
| Leonard St., 71 | | | S. A. Warner | 1 | " |
| Do | 73 | | J. F. Duckworth | 1 | " |
| Do | 80 | Paton & Co | J. H. Giles | 1 | " |
| Do | 84 | Mr. Sniffin | | 1 | " and Basement. |
| Liberty St., 25, 27 | | W. B. Windle | | 1 | " |
| Do | 29 | F. W. Lasak | | 1 | " |
| Do | 85 | | | 1 | " |
| Do | 93 | J. M. Matthews | | 1 | " |
| Do | 95, 97 | Murphy & Benedict | | 2 | " |
| Do | 96 | Mr. McBride | | 1 | " |
| Do | 99 | A. R. Eno | | 1 | " |
| Do | 103 | T. Strang | | 1 | " |
| Do | 105 | J. J. Henry | | 1 | " |
| Do | | | J. B. Snook | 1 | " and Basement. |
| Ludlow St., near Houston | | | H. Fernback | 4 | " |
| Do | | B. R. Winthrop | | 1 | " |
| Madison Av. | | | | | Columns—Ch. of Incarnation. |
| Maiden Lane 2 | | W. H. Smith | | | 1 Store Front. |
| Do | 4 | Mr. Young | | 1 | " |
| Do | 6 | W. H. Smith | | 1 | " |
| Do | 8 | | | 1 | " |
| Do | 9, 11, 13, | Swan & Co | | 3 | " |
| Do | 10 | | | 1 | " |
| Do | 15 and 25 | W. H. Smith | | 2 | " |
| Do | 17, 19 | J. Fellowes & H. Young | | 2 | " and Rears. |
| Do | 21, 23 | Fellowes & Schell | | 2 | " |
| Do | 22 | J. E. Hyde's Sons | | 1 | " |
| Do | 31 | | | 1 | " |
| Do | 33 | P. Murray | | 1 | " |

| LOCATION. | | PROPRIETOR. | ARCHITECT. | DESCRIPTION. |
|---|---|---|---|---|
| NEW YORK: | | | | |
| MaidenLane, | 35 | L. Murray | | 1 Store Front. |
| Do | 38 | | | 1 " |
| Do | 47 | | | 1 " |
| Do | 51 to 61 | A. H. Wood | | 5 Store Fronts. |
| Do | 56 | W. B. Windle | J. B. Snook | 1 " |
| Do | 58 | | | 1 " and Basement. |
| Do | 63 | R. Mortimer | | 1 " |
| Do | 123 | R. & N. Dart | | 1 " |
| Do | cor. Little Green. | Platt Bros | | 1 " |
| Do | " " | W. H. Smith | | 3 " |
| Do | " Nassau | Est. of J. Duidam | | 39 feet " |
| Mercer St., | 5, 7 | | J. B. Snook | 2 Store Fronts. |
| Do | 18 | A. T. Stewart | Kellum & Son | 5-story Front and Basement. |
| Do | cor Howard | A. R. Eno | | 1 Store Front. |
| Do | Rear 555 B'way. | John Taylor | | 1 Store Rear. |
| Do | | W. Gibson | | 1 Store Front. |
| Mott St., | | I. M. Singer & Co | G. H. Johnson | Sewing Machine Manfy. Pl. IV. and VI. |
| Murray St., | 6 | | | 1 Store Front. |
| Do | 8, 10 | E. Parmly | | 2 " |
| Do | 14 | J. L. Platt | | 1 " |
| Do | 16 | O. Thompson | | 1 " |
| Do | 17 to 29 | H. D. Aldrich | S. A. Warner | 7 " |
| Do | 36 | Dr. Scott | | 1 " |
| Do | 37, 39 | A. Higgins | | 2 " |
| Do | 41 | | Thomas & Son | 1 " |
| Do | 45 | Mr. Hutchings | S. A. Warner | 1 " |
| Do | 46 | | S. A. Warner | 1 " |
| Do | 47 | W. Sturtevant | J. C. Wells | 1 " |
| Do | 49 | | | 1 " |
| Do | 55 | | | 1 " |
| Do | cor College Pl. | Mr. Stevens | | 189 ft. " |
| Do | | | | 1 Store Front. |
| Do | | Dr. Hunter | | 1 " |
| Nassau St., | 33 | | | 1 " |
| Do | 52, 54 | C. & U. J. Smith | | 2 " |
| Do | 115 to 121 | N. C. Platt | J. Sexton | 4 " and Rears, Pl. XXXV. No.34 |
| Do | cor Maiden Lane. | Mr. Swan | | 2 " |
| Do | | Mr. Taylor | | 1 " |
| Do | | Mr. Youngs | | 1 " |
| New William St., | 10 | W. H. Smith | | 1 " |
| Park Place | 3 | F. Pares | King & Kellum | 1 Store Front and Basement. |
| Do | 9, 11 | E. Parmly | | 2 " |
| Do | 12 | Judge Roosevelt | | 1 " |
| Do | 17 | J. L. Platt | | 1 " |
| Do | 19 | O. Thompson | | 1 " |
| Do | 21 | E. B. Strange | | 1 " |
| Do | Church & Barclay. | C. W. & J. T. Moore | Thomas & Son | 147 ft. " |
| Do | College Pl. & Barclay | Chittenden, Bliss & Co. | S. A. Warner | 138 ft. " |
| Do | and Murray | Lathrop & Ludington | " | 117 ft. " |
| Do | cor. Church | Wm. Watson | | 167 ft. " |
| Do | " | W. G. Hunt & Co | King & Kellum | 145 ft. " |
| Do | " College Pl. | Thomas Hunt | " | 150 ft. " |
| Do | | E. Parmly | | 1 Store Front. |
| Do | | D. B. St. John | | 2 " |
| Do | | Christie & Constant | | 1 " and Basement. |

| LOCATION. | | PROPRIETOR. | ARCHITECT. | DESCRIPTION. |
|---|---|---|---|---|
| **NEW YORK:** | | | | |
| Park Place | | T. Slocum | | 100 ft. Front. |
| Do | | Spofford & Tileston | | 2 Store " |
| Park Row, | 13, 15 | Mr. Bangs | | 2 " |
| Do | | W. B. Astor & J. J. Phelps. | Thomas & Son | 5 " |
| Pearl St., | 282 | H. V. Hendrick | | 1 " |
| Do | | W. H. Cary | King & Kellum | 2 " |
| Do | | Mary Chesebrough | " | 1 " and Basement. |
| Do | | M. Halsted | " | 1 " |
| Do | | J. H. Coster | " | 1 " |
| Do | | R. Carmley | " | 1 " |
| Do | | P. Williams | " | 1 " |
| Do | | L. G. Morris | " | 1 " |
| Do | | | " | 38 ft. " |
| Do | cor. Moore | | | 1 Store Front. |
| Peck Slip, | " Front | J. S. Harris & Co. | R. G. Hatfield | 103 ft. Front. |
| Pine St., | 11 | A. J. Cipriant | | 2-story Front. |
| Reade St., | 74 | Mr. Bradshaw | | 1 Store Front. |
| Do | 97, 99, 101 | A. Higgins | J. F. Duckworth | 3 " |
| Do | 103 | Gilbert Estate | | 1 " and Rear on Chambers St. |
| Do | 104 | | | 1 " |
| Do | cor. Church | Read & Bradshaw | G. H. Giles | 86 ft. " and Basement. |
| Do | " " | | Kellum & Son | 160 ft. " " |
| Do | | Bliss, Briggs & Douglas. | S. A. Warner | 100 ft. " |
| Do | | R. H. McCurdy | | 50 ft. " and basement. |
| Do | | J. Q. Jones | | 5-story " Pl. XV., No. 8. |
| Sixth Avenue, No. 206 | | G. P. Rogers | | 1 Store Front. |
| Third " 805 | | | W. T. Beers | 1 " |
| Third Ave., cor. 34th St | | H. Hughes | do | 56 feet Store Front. |
| Third and Fourth Av., Astor | | | | [Inside Cast Iron Work. |
| Place and Seventh St | | Peter Cooper | F. A. Petersen | 326 ft. Store Front, Cooper Inst., and all |
| Twenty-fourth Street | | A. R. Eno | | Fifth Avenue Hotel Front. |
| University Place | | Society Library | Thomas & Son | Main Cornice & Balust'de. Pl. XX., No. 93. |
| Vesey St., No. 22 | | W. Morris | | 1 Store Front. |
| Do | 36 | S. Sutton | | 1 " |
| Do | 45 | J. Osborn | | 1 " |
| Do | | S. Sutton | | 50 ft. Store F'nt, 5-story, sim. to Pl. LVIII. |
| Walker St., No. 24 | | M. H. Litchstein | King & Kellum | 1 Store Front and Basement. |
| Do | 36 | T. Lewis | | 1 " |
| Do | 38 | Mr. Lewis | | 1 " |
| Do | 44 | G. Johnson | Thomas & Son | 1 " |
| Do | 48, 50 | Mr. Lane | R. G. Hatfield | 2 " [XV., No. 7 |
| Do | 61 | Mrs. Goelet | | 5-story Front and Basement, similar to Pl. |
| Do | | J. Lee | Kellum & Son | 5 " " " sim. to Pl. CII. |
| Wall St., Nos. 8 to 20 | | J. G. Pearson and others. | J. G. Pearson | 6 Store Fronts, 2-story, similar to Plate XXXVII., No. 17. |
| Do | 49 | | | 1 Store Front. |
| Do | | Mechanics' Bank | R. Upjohn & Co. | Dome and Lantern, 126 feet circumference. |
| Warren St., Nos. 4, 6 | | S. V. Hoffman | | 2 Store Fronts. |
| Do | 8 | A. M. Lyon | | 1 " |
| Do | 11 | J. Lee | R. G. Hatfield | 1 " |
| Do | 12, 14 | A. Higgins | | 2 " |
| Do | 15 | F. E. Gilbert | | 1 " |
| Do | 16 | T. March | | 1 " |
| Do | 17, 19 | Henrys, Smith & Townsend | | 2 " |
| Do | 18, 20 | Mr. Cleveland | | 2 " |

| LOCATION. | PROPRIETOR. | ARCHITECT. | DESCRIPTION. |
|---|---|---|---|
| **NEW YORK:** | | | |
| Warren Street, 23, 25 | C. A. Bandouine | Mr. Gardiner | 51 ft. Store Front. |
| Do 24 | Mr. Martin | | 1 " |
| Do 26, 28, 30 | Allan, McComb & Langlois | | 3 " |
| Do 37, 39 | T. U. Smith & J. J. Henry | | 2 " |
| Do 38 | | | 1 " |
| Do 40 | C. A. Bandouine | Mr. Gardiner | 1 " |
| Do 41, 43 | T. Suffern | J. W. Ritch | 2 " |
| Do 42 | J. A. Stevens | | 1 " |
| Do 49 | Rogers & Walker | S. A. Warner | 1 " |
| Do 51 | Mr. Center | | 1 " |
| Do 53 | F. E. Gilbert | | 1 " |
| Do 55 | Judge Whiting | | 1 " |
| Do 76 | Mrs. P. Bonnet | Thomas & Son | 1 " |
| Do cor. Church | H. D. Aldrich | S. A. Warner | 2 " |
| Water St., No. 120 | A. Hendricks | | 1 " |
| White do 79 | S. Kohnstamm | Thomas & Son | 1 "   [Pl. CII. |
| Do cor. Franklin Pl. | S. H. & J. E. Condict | Kellum & Son | 73 ft. "   5-story, Basement and Rear, |
| Whitehall St. | Union Ferry Co. | J. Kellum | South and Hamilton Ferry Houses. |
| Do | Corn Exchange | E. L. Roberts | Iron Work Corn Exchange. |
| William St., No. 93 | A. H. Ward | | 1 Store Front. |
| Do 128 | | | 1 " |
| Do 130 | A. B. &. D. Sands | | 1 " |
| Do 163 | B. A. Field | | 1 " |
| Do cor. Ann | S. N. Livingston | | 1 " |
| Do | Great Western Ins. Co. | Renwick & Sands | Insurance Building. |
| Worth St. | | S. A. Warner | 315 feet Store Front and Basement. |
| Do No. 39 | | J. F. Duckworth | 1 Store Front. |
| Do 41 | | J. F. Duckworth | " |
| Do 43, 45 | Mr. Nesmith | S. A. Warner | 50 ft. " |
| OSWEGO, N. Y | Oswego Hotel Co. | W. T. Beers | 300 feet Front. |
| Do | T. Kingsford & Son | | Oswego Starch Factory. |
| PANAMA | Panama Railroad Co. | | Railroad Depot. |
| Do | do | | Verandah, &c. |
| PETERBORO', C. W | R. Nicholls | | 2 Store Fronts.   Plate XXVIII., No. 65. |
| PETERSBURG, Va. | D. A. Paul | | 1 Store Front. |
| Do | Lyon, Abraham & Davis | | 1 " |
| **PHILADELPHIA:** | | | |
| Arch St., No. 116 | Jones, White & Oo. | | 1 " |
| Do 124 | W. H. Hart | | 1 " |
| Chestnut St., 49 | A. Masson | | 1 " |
| Do 51 | do | | 1 " |
| Do 52 | J. A. Gendell | | 1 " |
| Do 54 | W. W. Keen | | 1 " |
| Do 56 | J. A. Gendell | | 1 " |
| Do 61 | do | | 1 " |
| Do 63 | do | | 1 " |
| Do 65 | Mr. Landreth | | 1 " |
| Do 85 | Mr. Lewis | | 1 " |
| Do 87 | do | | 1 " |
| Do 115 | W. W. Keen | | 1 " |
| Do 123 | S. H. Hoxie | | 1 " |
| Do 136 | Bailey & Co. | | 1 " |
| Do | J. F. Fisher | | 3 Store Fronts. |
| Do | Dr. Swaim | | 3 "   Swaim's Building. |
| Do | Jules Harrel | | 1 Store Front. |

| LOCATION. | PROPRIETOR. | ARCHITECT. | DESCRIPTION. |
|---|---|---|---|
| **PHILADELPHIA :** | | | |
| Chestnut St ........... | Mr. Dunbar........... | ............ | 2 Store Fronts.  Pl. LXIV., No. 25. |
| Do ........... | Girard Estate......... | ............ | 2 " Girard Buildings. |
| Do ........... | Williamson & Mellor... | ............ | 3 " Pl. LXIV., No. 25. |
| Do ........... | A. Foit............. | | 1 Store Front. |
| Commerce St......... | C. C. Cope........... | | 1 " |
| Lodge Alley ......... | Williamson & Mellor... | ............ | 3 Store Fronts. |
| Do ......... | Mr. Dunbar........... | | 2 " |
| Market St., No.  81.... | C. H. Fisher......... | | 1 Store Front. |
| Do        147.... | M. L. Hollowell........ | ........ | Rolling Shutters for Front. |
| North Third St........ | Mr. Madora............ | | 1 Store Front. |
| Sixth St............. | D. Landrith ......... | | 2 Store Fronts. |
| South Fourth St....... | R. T. Shepherd........ | | 2 " |
| Do ....... | W. Ford ............ | | 1 Store Front. |
| Do ....... | J. A. Gendell......... | | 3 Store Fronts. |
| Strawberry St......... | W. W. Keen........... | | 2 " |
| Third St............. | Mr. Ballinger ........ | | 2 " Pl. XXXV., No. 34. |
| Do ............. | Mr. Fassit ........... | | 1 Store Front. |
| Do ............. | Siegur, Lamb & Co..... | | 1 " |
| Do ............. | Faust & Wineburne.... | | 1 " |
| Do ............. | Mr. Stone ............ | | 1 " |
| Do ............. | Towns & Sharpless..... | | 1 " 2 stories. |
| Washington Ave....... | Penn. Railroad Co...... | G. H. Johnson..... | Grain Warehouse.  Plates LX., LXI., LXII. |
| | | | 107 × 125 ft.   5 stories.   Fire-proof. |
| **PITTSBURG, Penn**........... | Mr. Yeager .......... | | 1 Store Front. |
| Do ........... | A. A. Mason & Co..... | | 1 " |
| Do ........... | C. H. Paulson......... | | 2 Store Fronts. |
| Do ........... | J. Brown............. | | 2 " sim. to Pl. XXXV., No. 34. |
| **PITTSFIELD, Mass**........... | J. C. West........... | | 3 " |
| Do ........... | Plumpkit & Hartbut ... | | 1 Store Front. |
| **PORTLAND, Me**......... | H. N. Jose........... | | 2 Store Fronts. |
| **PROVIDENCE, R. I**......... | G. A. Howard......... | | 475 feet Store Fronts. |
| Do ......... | Tolman & Bucklin..... | | 5 Store Fronts. |
| Do ......... | Wm. Andrews......... | | 2 " |
| Do ......... | Mr. Duncan.......... | | 8 " |
| Do ......... | S. Dexter............ | | 2 " |
| Do ......... | B. D. Wheedon........ | | 2 " |
| Do ......... | H. Rogers............ | | 2 " |
| Do ......... | J. Arnold............ | | 4 " |
| **RICHMOND, Va**........... | Stebbing, Darricott & Co... | | 3 " |
| Do ........... | W. Barrett........... | | 5 " |
| Do ........... | Kent, Payne & Kent ... | | 2 " |
| Do ........... | J. P. Ballard......... | | 1 Store Front. |
| Do ........... | O. A. Stryker........ | | Rolling Shutters for Front. |
| **RIO JANEIRO, Brazil**....... | Dr. T. Rainey ........ | J. Whyte.......... | Ferry Ho., 100 ft. Front.  Pl. LXXXVIII. |
| **ROCHESTER, N. Y**......... | Elwanger & Barry ..... | | 75 feet Store Front. |
| Do ......... | Mr. Erricson.......... | | 86 " " |
| Do ......... | W. A. Reynolds........ | | 58 " " |
| Do ......... | Samuel Wilder ....... | | 1 Store Front. |
| Do ......... | D. W. Powers ........ | | 1 " |
| **ROME, N. Y**............. | Hill, Brothers & Co.... | | 1 " |
| **SACRAMENTO, Cal**......... | | | 1 " |
| **SAN FRANCISCO, Cal**....... | R. M. Sherman........ | | 2 Store Fronts. |
| Do ...... | G. R. Jackson & Co..... | | 1 Store Front. |
| Do ...... | J. B. Snook .......... | J. B. Snook....... | 2 Store Fronts. |
| Do ...... | | | 118 feet Store Front. |

| LOCATION. | PROPRIETOR. | ARCHITECT. | DESCRIPTION. |
|---|---|---|---|
| SAN FRANCISCO, Cal. | | | 34 feet Store Front. |
| SAVANNAH, Ga. | S. C. Deming | | 2 Store Fronts. |
| SCRANTON, Penn. | Mr. Shopland | | 1 Store Front. |
| SHARON SPRINGS, N. Y. | | L. Burgher | Pavillion over Spring. |
| SPRINGFIELD, Mass. | Foot & Co. | | 4 Store Fronts. |
| Do | D. W. Barnes | A. L. Chapin | 1 Store Front. |
| Do | John Madden | | 1 " |
| STAMFORD, Conn. | Augustus Weed | M. B. Wolsey | 38 feet Store Front and Balcony. |
| SYRACUSE, N. Y. | Dillaye, Brothers | | 3 Store Fronts. |
| TROY, N. Y. | H. E. & W. Allendorph. | | 1 Store Front. |
| Do | Jacobs & Caswell | | 1 " |
| Do | Troy City Bank | | 38 feet Store Front. |
| Do | L. Smith | King & Tecknitz | 1 Store Front. |
| TROY, Penn. | E. W. Perrine | | 1 " |
| UTICA, N. Y. | J. Sayer | | 1 " |
| VICKSBURG, Miss. | J. B. Wheeler & Co. | | 1 " [XII. |
| WATERVLIET, N. Y. | U. S. Government | | Arsenal Storehouse, 100 by 196 feet. Pl. |
| Do | do | | Arsenal. Iron Work. |
| WASHINGTON, D. C. | do | | Extension Congressional Library. |
| Do | do | | Patent Iron Lathing Extension, Treasury |
| Do | do | | Iron Work, Ford's Theatre. [Build'g |
| WILMINGTON, N. C. | W. A. Barry | | 1 Store Front. |

VIEW OF THE ARCHI

13TH & 14TH STS

ARCHITECTURAL IRON WORKS

RIVER, NEW YORK.

*Plate III*

*No. 30.*

ARCHITECTURAL IRON WORKS, — NEW-YORK

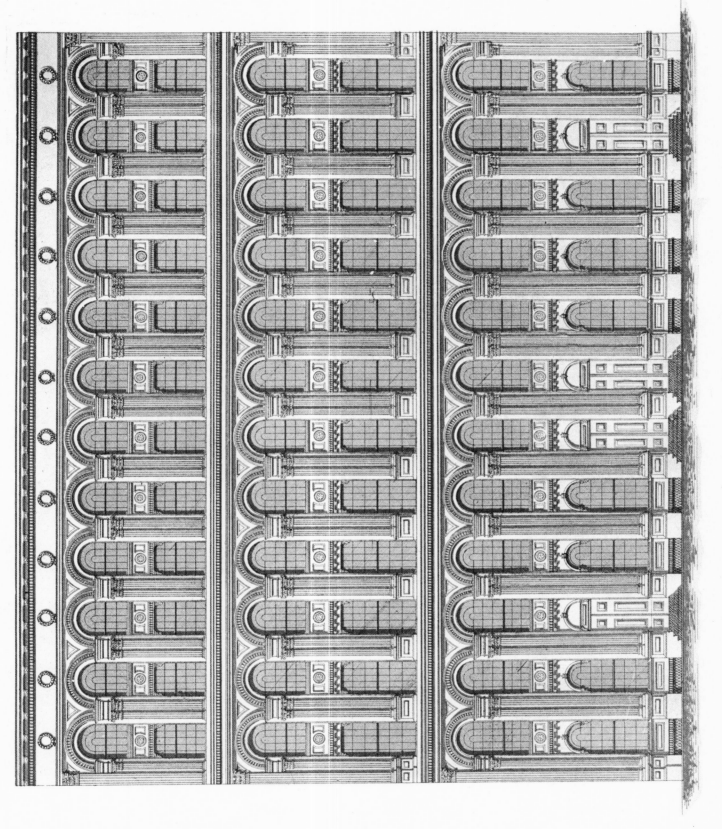

*Plate II*

*N° 5*

Front Elevation J. M. Singer & Co's Sewing Machine Manufactory

ARCHITECTURAL IRON WORKS, NEW YORK.

Scale—one inch to twelve feet

Plate V.

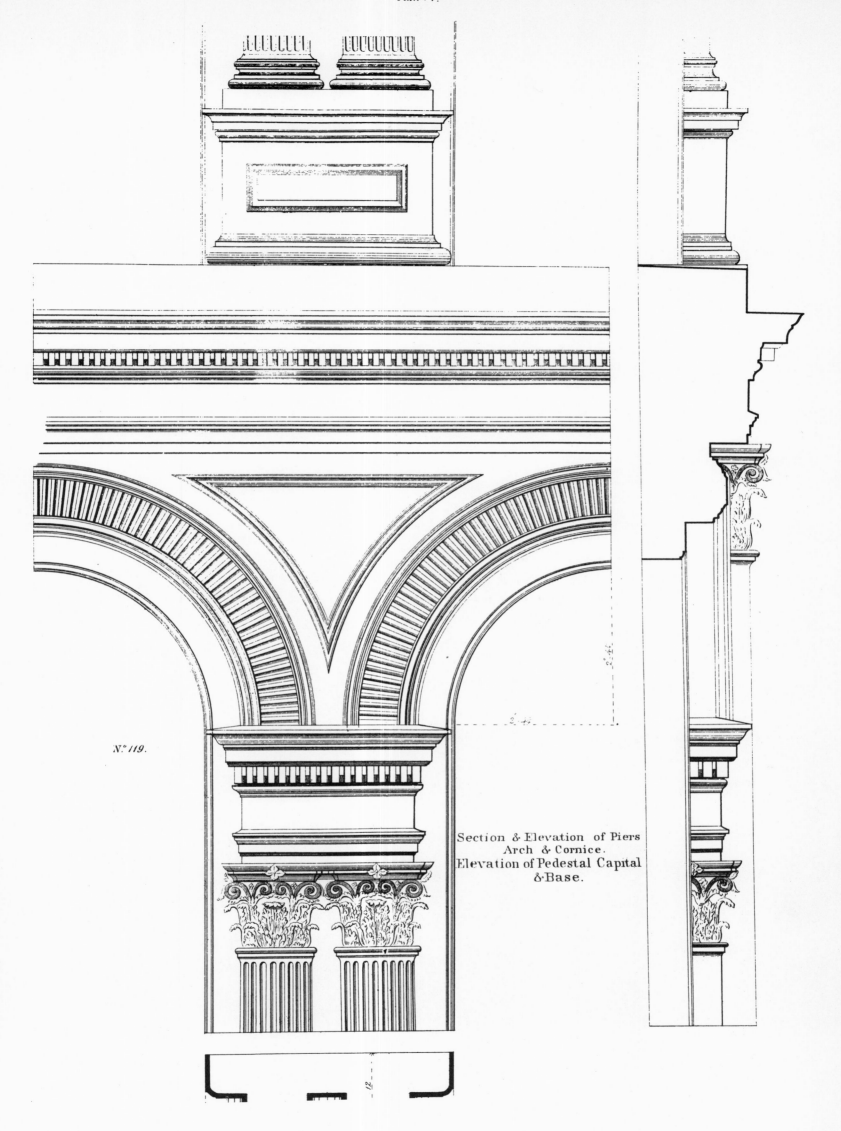

Nº 119.

Section & Elevation of Piers
Arch & Cornice.
Elevation of Pedestal Capital
& Base.

*Plate VI.*

## Section of Singer Building.

*N° 16*

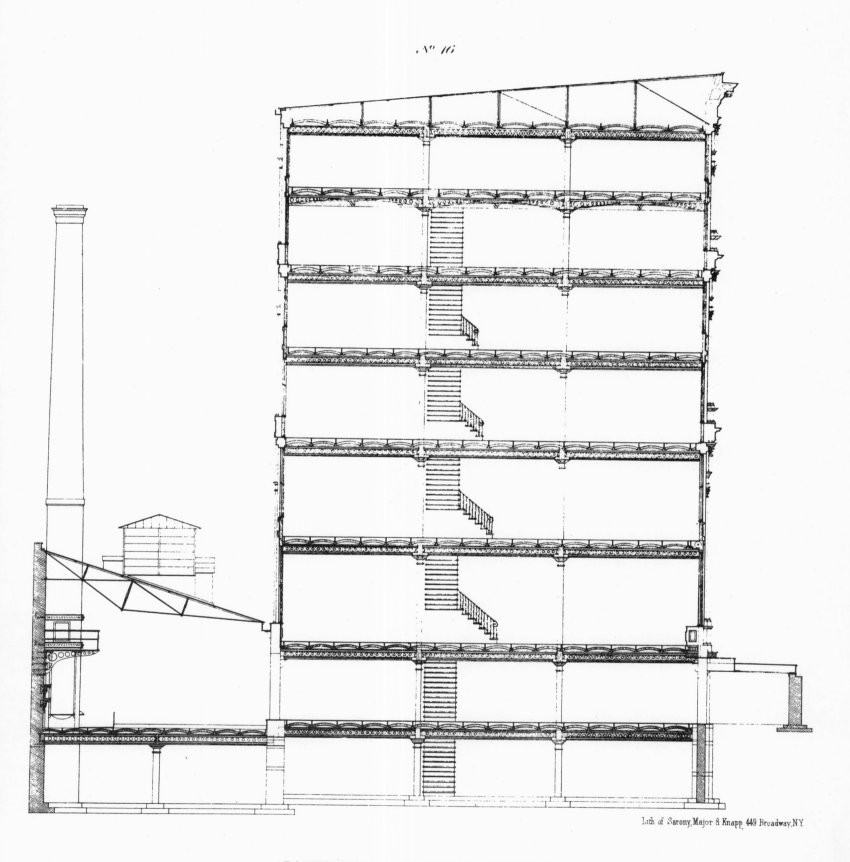

ARCHITECTURAL IRON WORKS,—NEW-YORK

*Plate VII.*

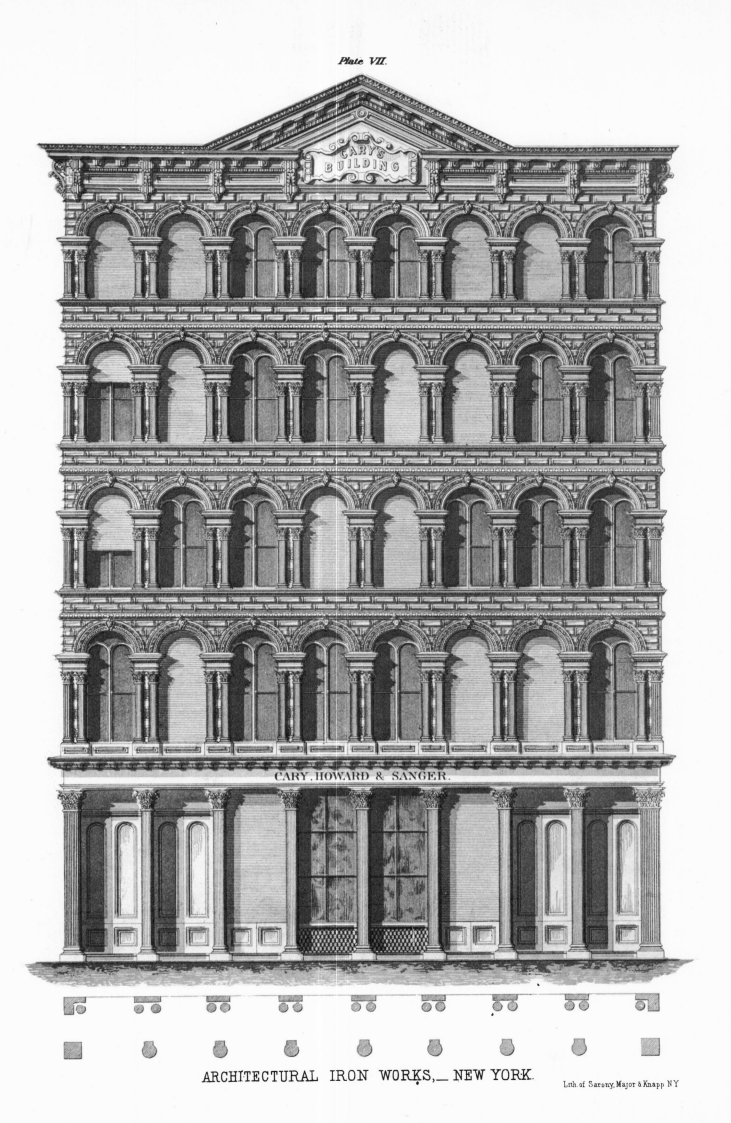

CARY'S
BUILDING

CARY, HOWARD & SANGER.

ARCHITECTURAL IRON WORKS,— NEW YORK.

Lith. of Sarony, Major & Knapp N.Y

Plate. VIII

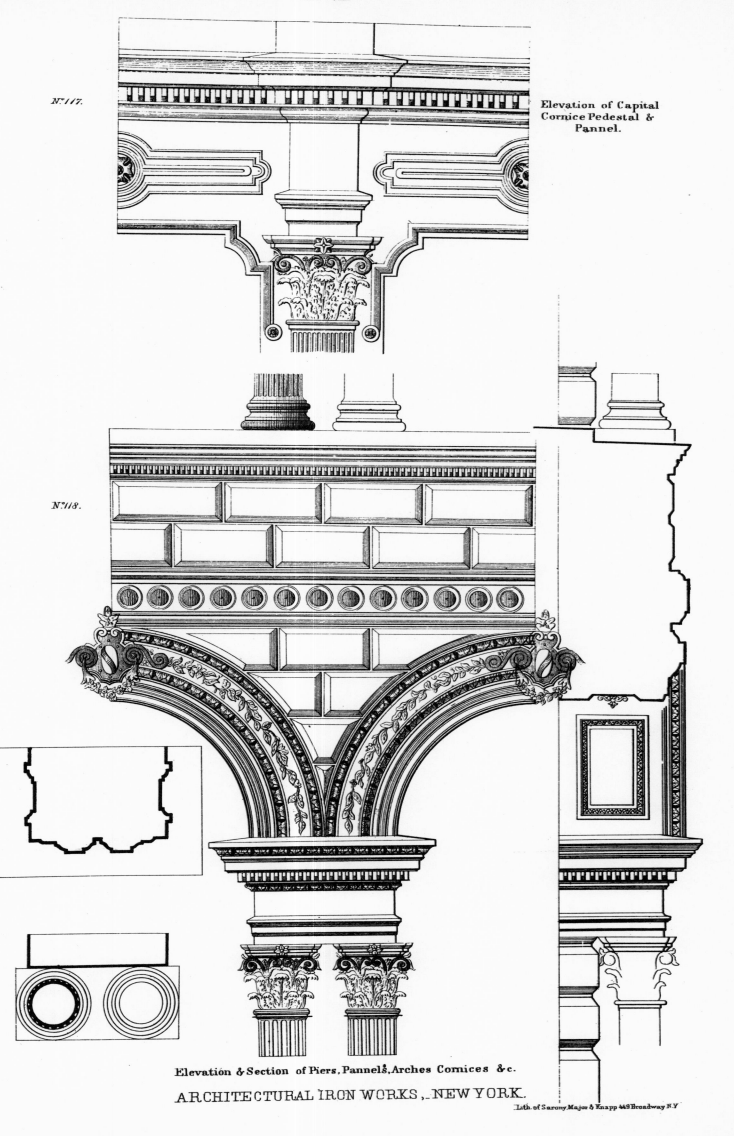

N.º 117.

Elevation of Capital
Cornice Pedestal &
Pannel.

N.º 118.

Elevation & Section of Piers, Pannels, Arches Cornices &c.

# ARCHITECTURAL IRON WORKS, NEW YORK.

Lith. of Sarony Major & Knapp 449 Broadway N.Y.

Plate IX.

No: 29.

ARCHITECTURAL IRON WORKS, — NEW-YORK.

*Plate X.*
# Cornices.

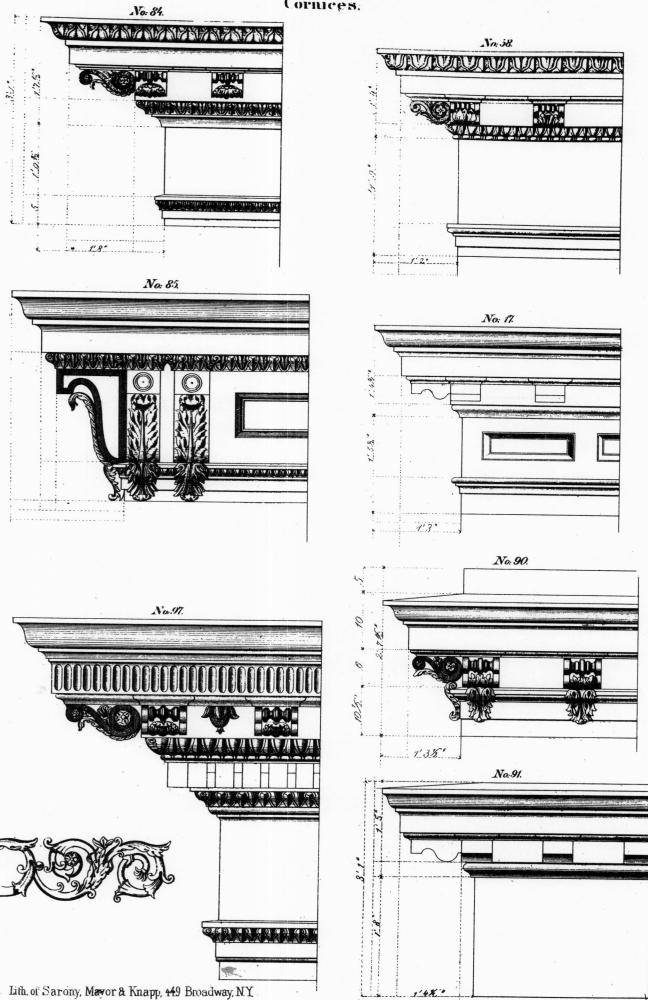

No: 84.

No: 38.

No: 85.

No: 17.

No: 97.

No: 90.

No: 91.

*Plate XI*
**OFFICE OF**

Total height 58' 6"

GROVER & BAKER S. M. Cº

**ARCHITECTURAL IRON WORKS _ NEW-YORK.**

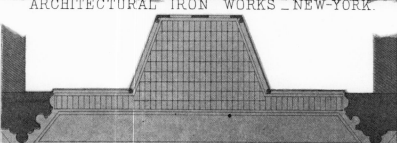

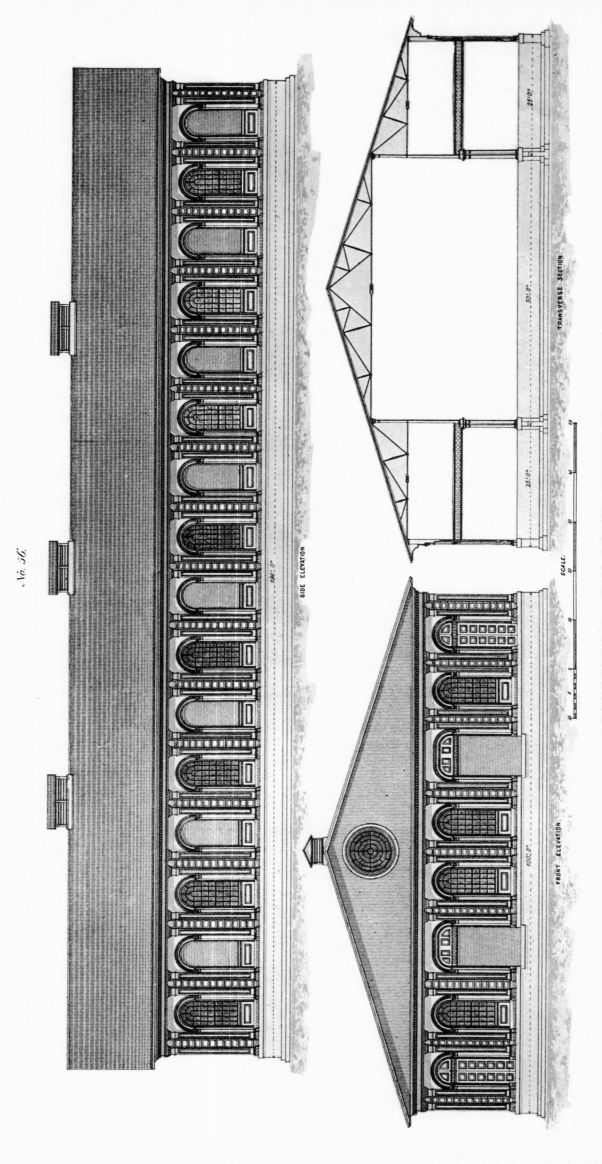

*Plate XII.*

Iron Store House for U. S. Arsenal,—Watervleit, N.Y.

No. 56.

SIDE ELEVATION

TRANSVERSE SECTION

FRONT ELEVATION

SCALE.

ARCHITECTURAL IRON WORKS,—NEW-YORK

Plate. XIII.

Section of Cornice and Arch.

Section and Elevation of Pier and Arch.

No. 120.

ARCHITECTURAL IRON WORKS,- NEW-YORK.

Lith of Sarony Major & Knapp 449 Broadway N.Y.

## ARCHITECTURAL IRON WORKS,—NEW-YORK.

*Plate XV*

## Designs for Store Fronts.

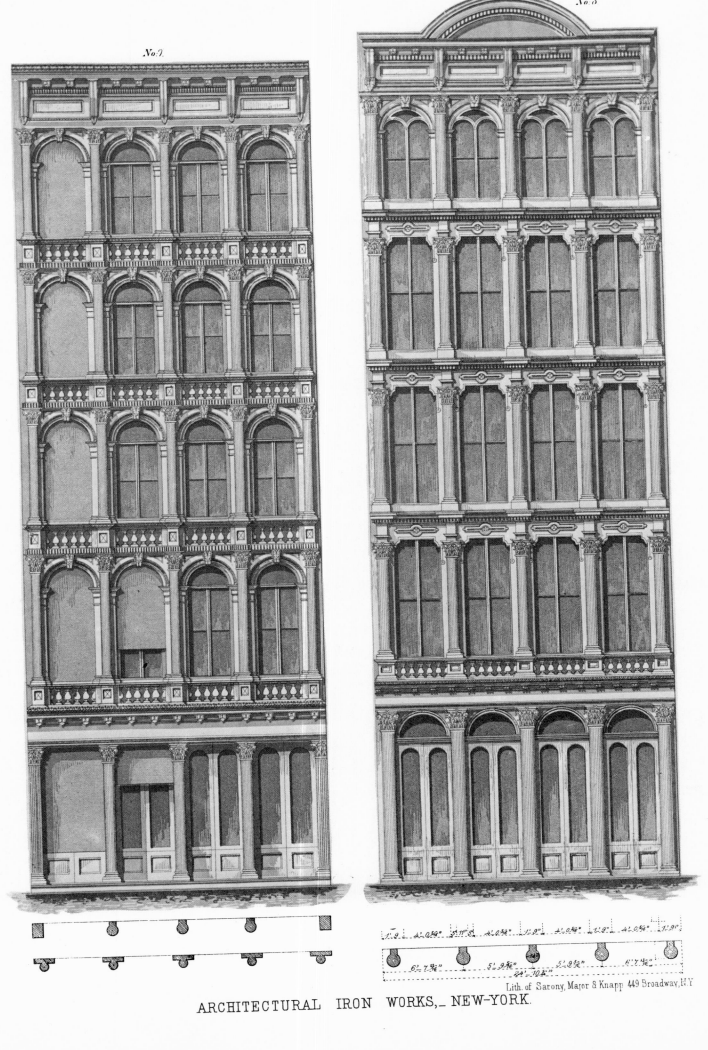

No. 7.

No. 8

ARCHITECTURAL IRON WORKS,— NEW-YORK.

Lith. of Sarony, Major & Knapp 449 Broadway, N.Y

*Plate XVI*

Cornices Balustrades and Pedestals.

No. 111.

No. 94.

No. 92.

No. 86

No. 82

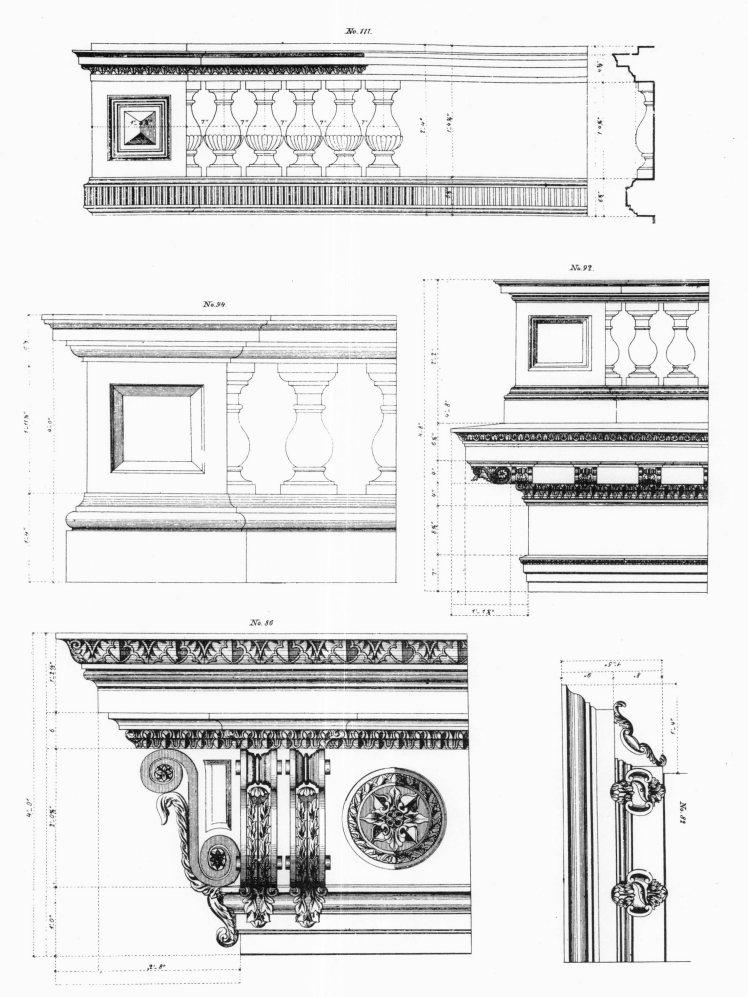

Lith of Sarony Major & Knapp. 449 Broadway NY.

ARCHITECTURAL IRON WORKS NEW YORK

*Plate XVII*

Designs for Store Fronts.

*No. 6.*

*No. 5.*

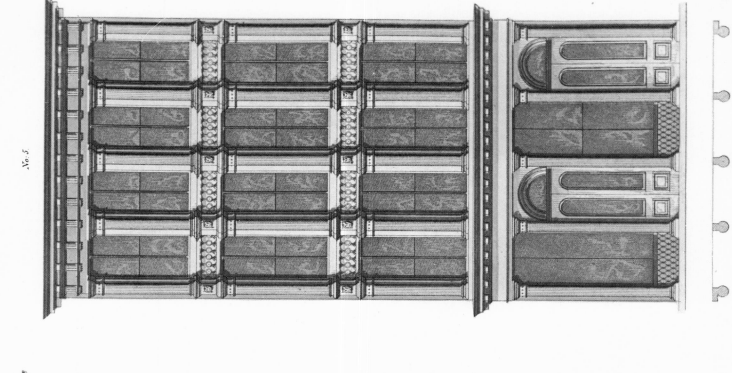

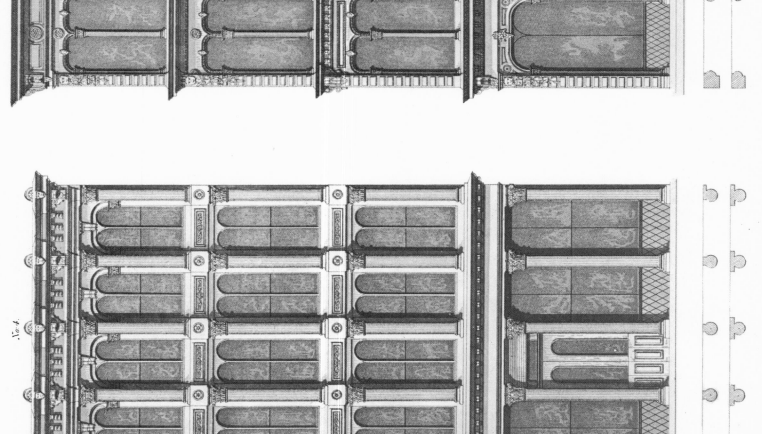

*No. 4.*

Lith. of Sarony Major & Knapp 449 Broadway N.Y.

ARCHITECTURAL IRON WORKS,— NEW-YORK.

Plate XVIII.
Cornices.

N.º 100.

N.º 103.

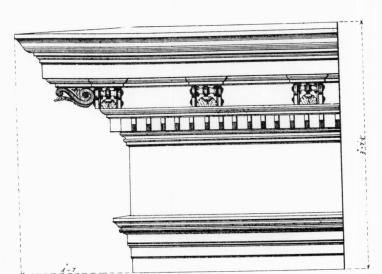

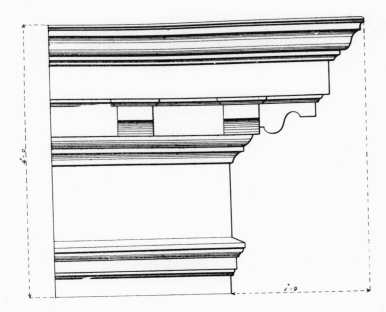

N.º 101.

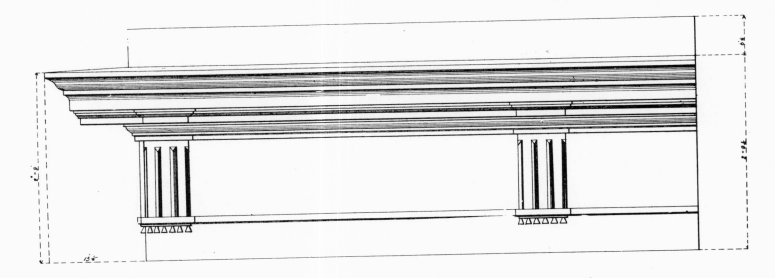

N.º 102.

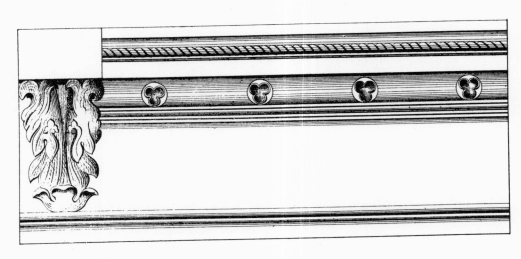

ARCHITECTURAL IRON WORKS. NEW YORK.

Lith of Sarony, Major & Knapp 449 Broadway N.Y.

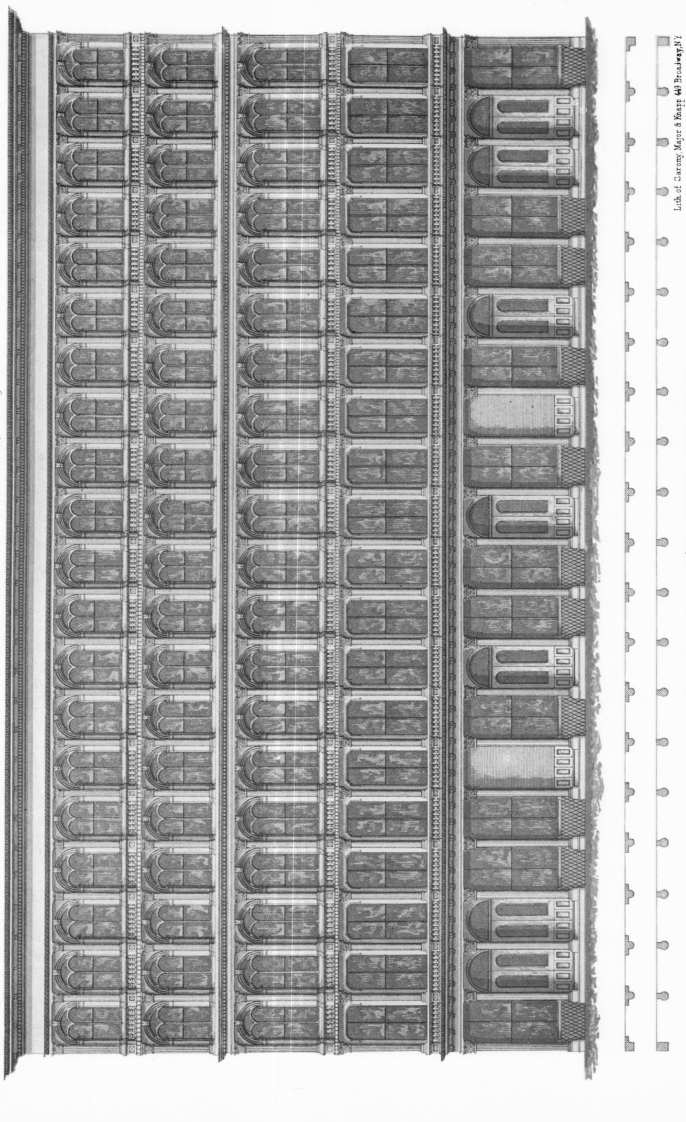

ARCHITECTURAL IRON WORKS,—NEW-YORK.

*Plate. XX.*

# Cornices.

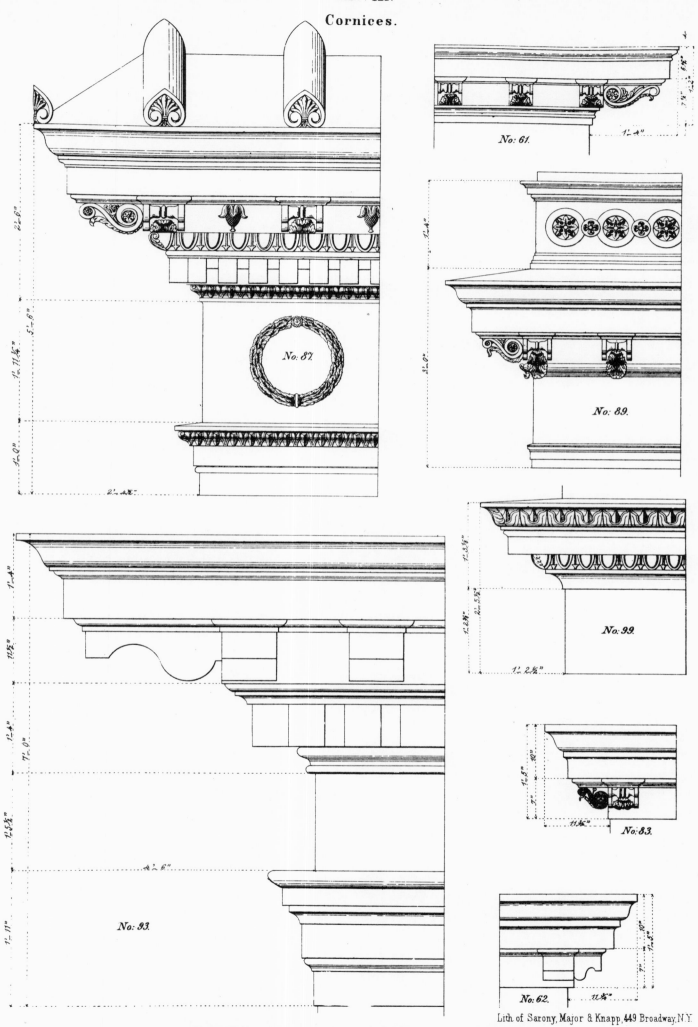

No: 61.

No: 87.

No: 89.

No: 99.

No: 83.

No: 93.

No: 62.

ARCHITECTURAL IRON WORKS,— NEW-YORK.

*Plate XXI.*

Design for Store Front.

*No 37.*

ARCHITECTURAL IRON WORKS,— NEW-YORK.

ARCHITECTURAL IRON WORKS  NEW YORK.

Lith. of Sarony, Major & Knapp. 449 Broadway N. Y.

*Plate XXIII.*

**Cornices.**

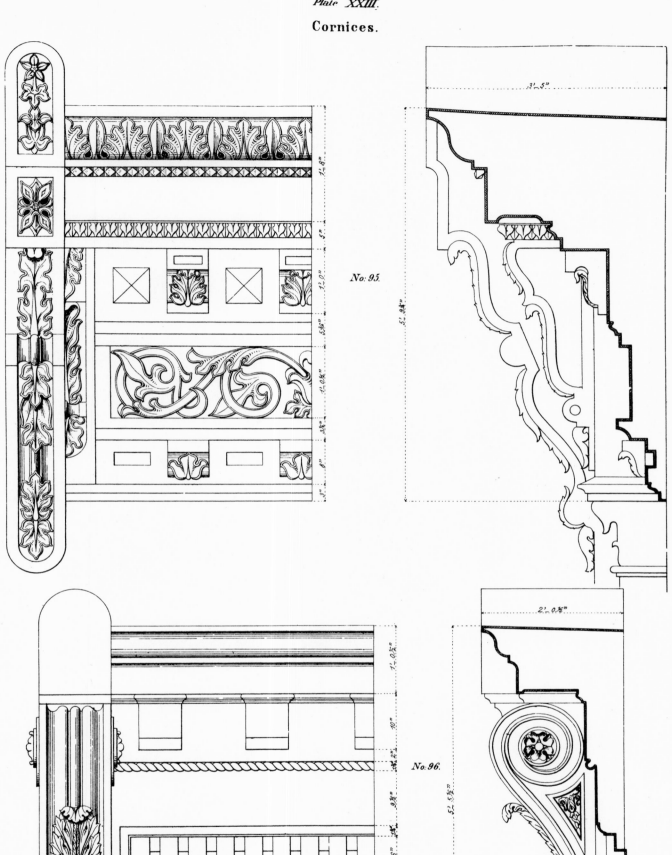

No: 95.

No: 96.

Lith. of Sarony, Major & Knapp, 449 Broadway, N.Y.

ARCHITECTURAL IRON WORKS,_ NEW-YORK

*Plate XXIV.*

## Front Elevation of first Story

*No.63.*

*No: 64.*

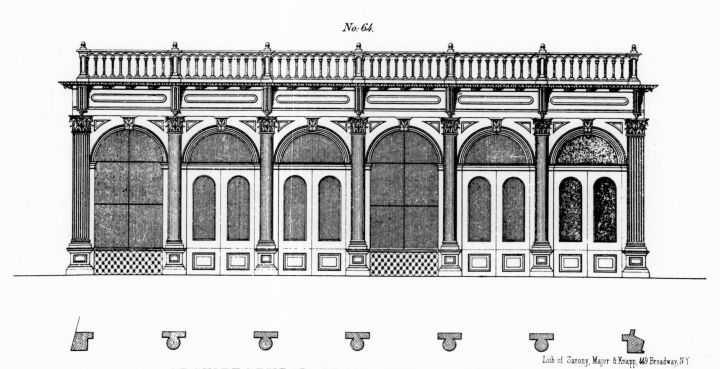

Lith of Sarony, Major & Knapp, 449 Broadway, N.Y.

ARCHITECTURAL IRON WORKS,___ NEW-YORK

*Plate. XXV.*

Arches and Tracery.     Capitals with Section of Piers.

*No.108.*

*No.109.*

*No.116.*

ARCHITECTURAL IRON WORKS,−NEW-YORK.

Lith. of Sarony, Major & Knapp, 449 Broadway, N Y

J. MᶜGREGOR.

ARCHITECTURAL IRON WORKS,—NEW-YORK.

Scale one Inch to Eight feet.

Lith. of Sarony Major & Knapp 449 Broadway N.Y.

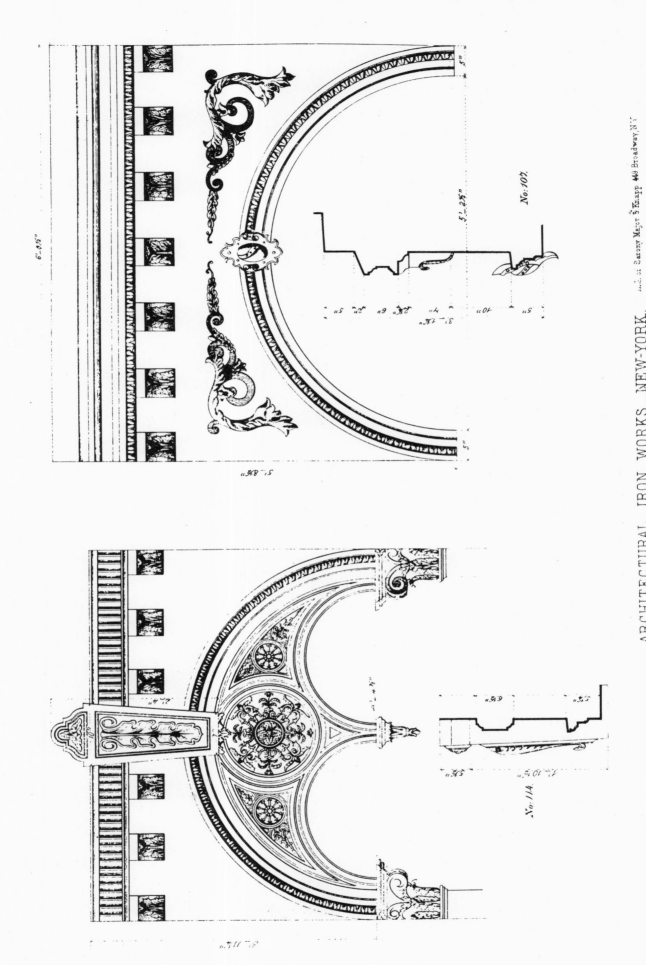

*Plate XXVII*

Arches, Keys, and Arch Ornaments.

ARCHITECTURAL IRON WORKS,—NEW-YORK.

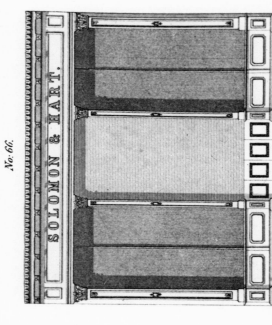

No. 66.

SOLOMON & HART.

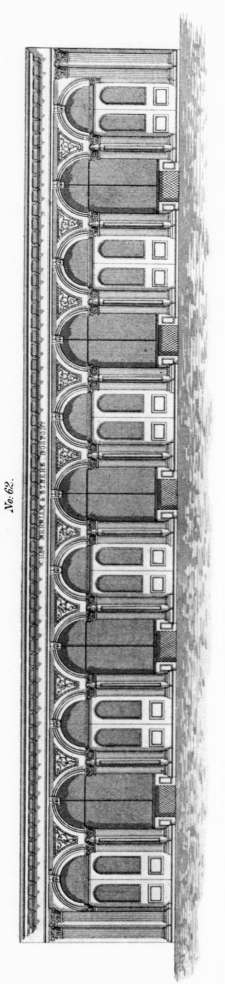

Plate XXVIII.

Designs for Store Fronts.

No. 62.

CHAS. MERRIAM & OTHERS, BOSTON

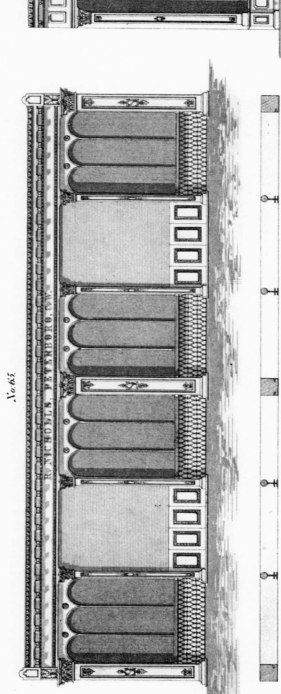

No. 65.

R. NICHOLLS, PETERBORO C.W.

Lith of Sarony, Major & Knapp 449 Broadway N.Y.

ARCHITECTURAL IRON WORKS,—NEW-YORK.

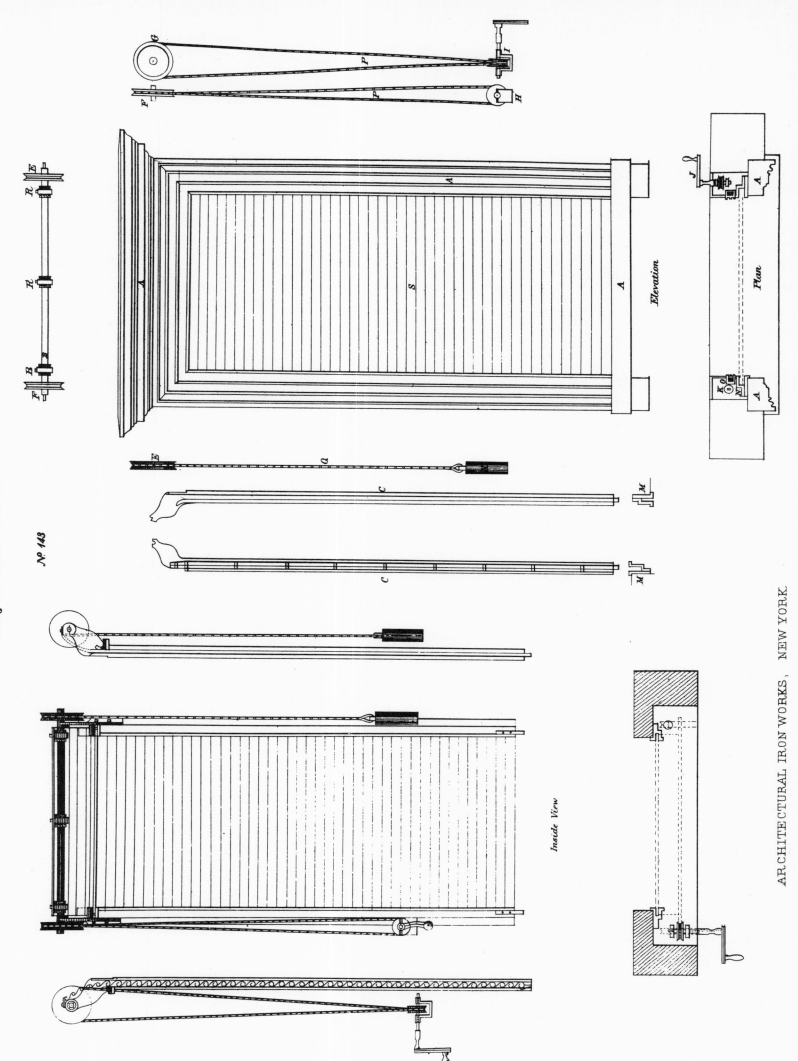

*Plate XXIX.*

Rolling Iron. Shutter & Fixtures.

Nº 143

*Elevation*

*Plan*

*Inside View*

ARCHITECTURAL IRON WORKS, NEW YORK.

Lith of Sarony Major & Knapp 449 Broadway NY.

*Plate XXX.*

*Nº 21.*

ODD FELLOWS HALL

ARCHITECTURAL IRON WORKS,— NEW-YORK.

*Plate XXXV*

Elevations of 1.ˢᵗ Story Fronts.

N.º 51.

N.º 52.

N.º 50.

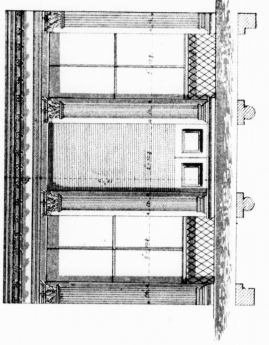

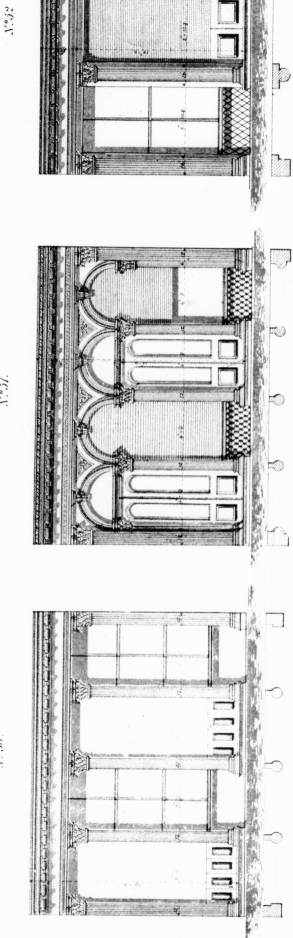

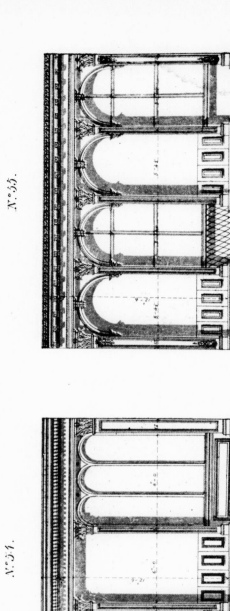

N.º 55.

N.º 54.

N.º 53.

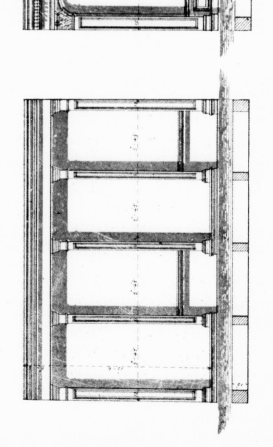

ARCHITECTURAL IRON WORKS. NEW YORK.

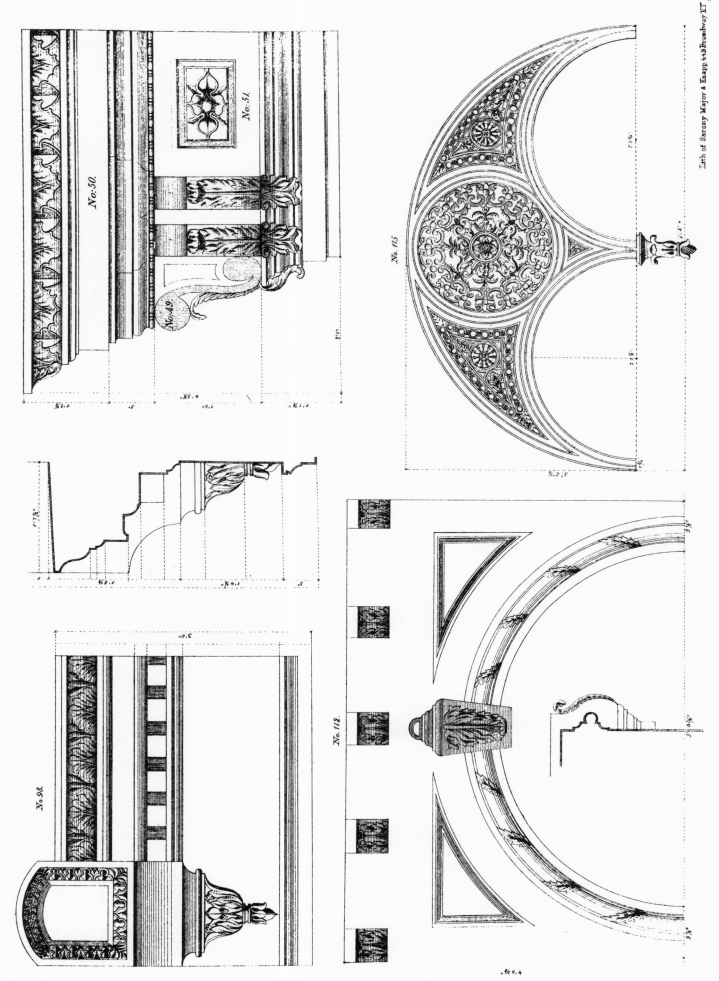

*Plate XXXII*

Cornices Arches & Arches Ornamental.

No. 50.

No. 51.

No. 49.

No. 98.

No. 112.

No. 115.

Lith of Sarony Major & Knapp 449 Broadway N.Y.

ARCHITECTURAL IRON WORKS NEW YORK

*Plate XXXIII*

Iron Store Front for W.B.Greenlaw & Co
Memphis    Tenn.

*No. 69.*

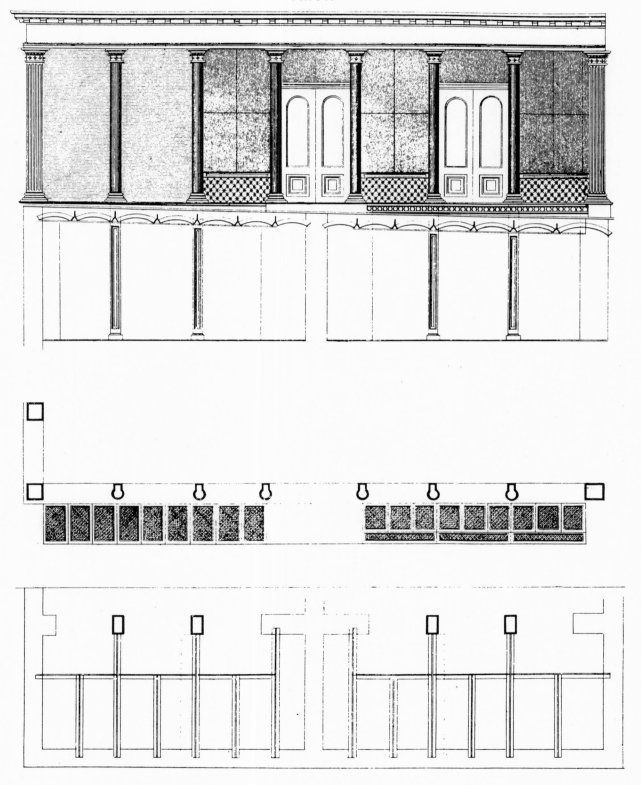

ARCHITECTURAL IRON WORKS.— NEW-YORK.

ARCHITECTURAL IRON WORKS,— NEW-YORK.

Lith of Carony Major & Knapp 449 Broadway N.Y.

*Plate XXVI*

*No. 33.*

*No. 34.*

ARCHITECTURAL IRON WORKS, NEW YORK

Lith of Sarony, Major & Knapp, 449 Broadway, N.Y.

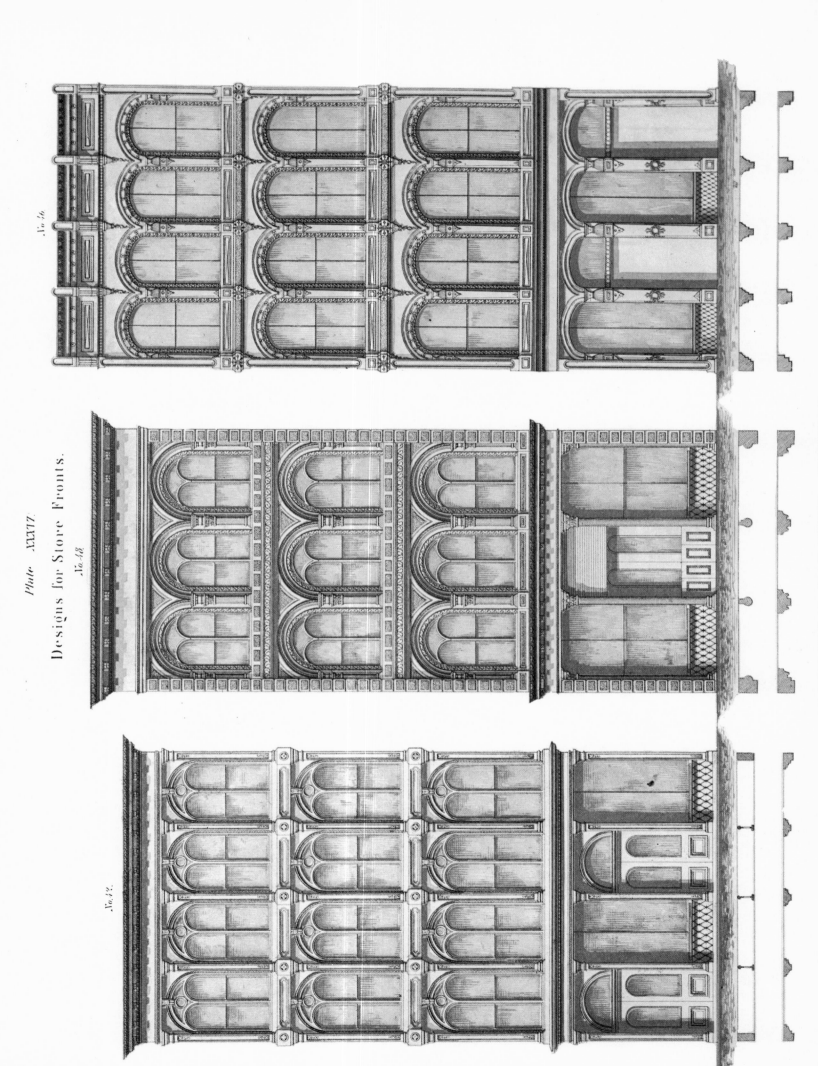

*Plate XXXVI.*

Designs for Store Fronts.

*No.48.*

*No.46.*

*No.47.*

ARCHITECTURAL IRON WORKS,– NEW-YORK.

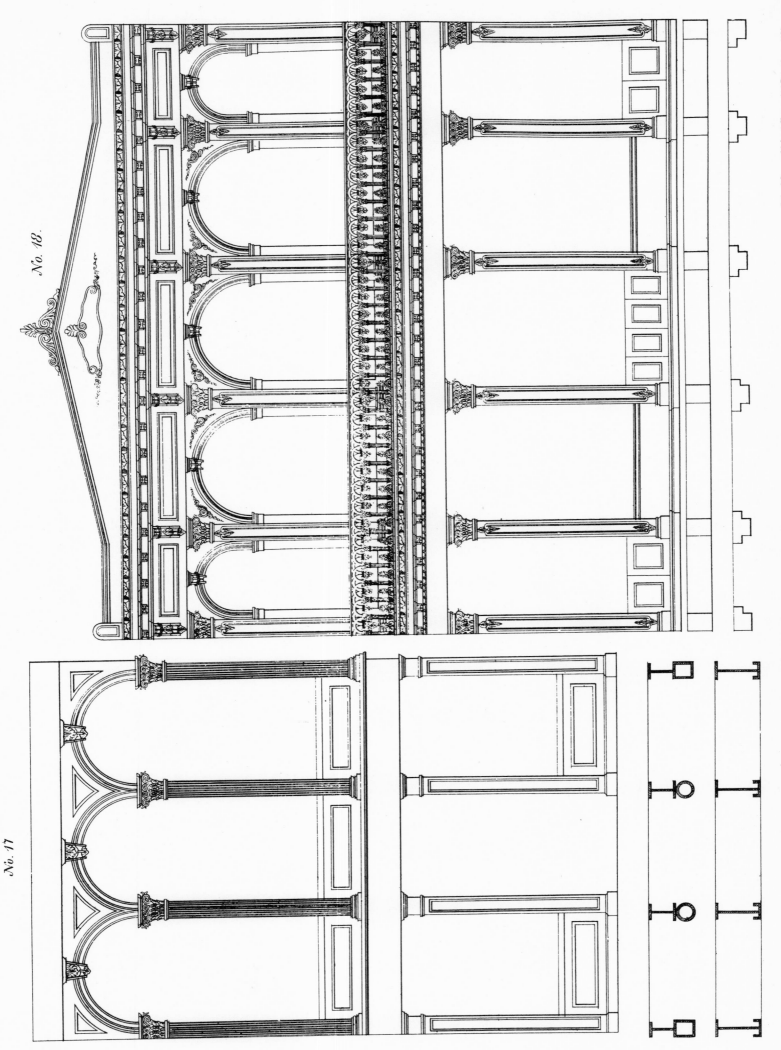

Plate XXXVII.

No. 18.

No. 17

ARCHITECTURAL IRON WORKS. NEW YORK.

Lith. of Sarony, Major & Knapp, 449 Broadway N.Y.

Plate XXVIII

ARCHITECTURAL IRON WORKS,— NEW-YORK

*Plate XXXIX*

Arches, Keys, Capitals & Section of Piers.

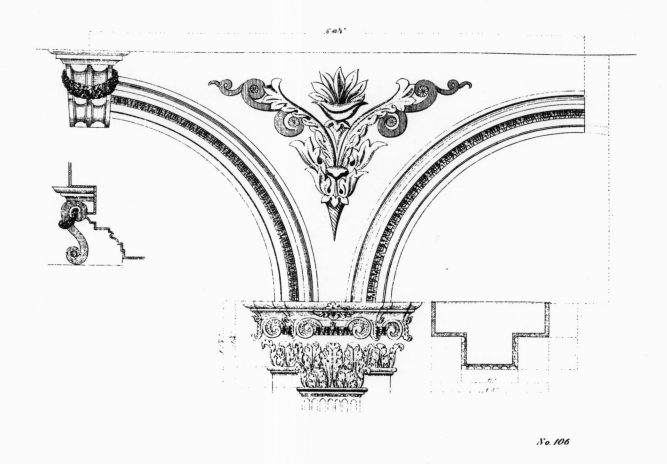

*No. 106*

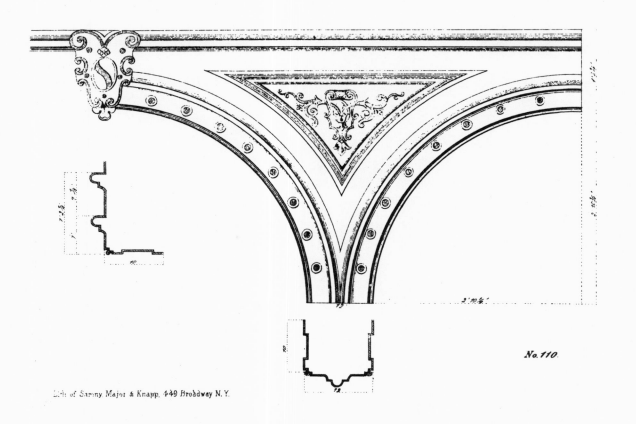

*No. 110.*

Lith of Sarony Major & Knapp, 449 Broadway N.Y.

ARCHITECTURAL IRON WORKS, NEW YORK.

*Plate XL.*

Store Front for M<sup>r</sup> Kramer
Boston, Mass          No:67.

Store Front for 267 Bowery
                              No:68.

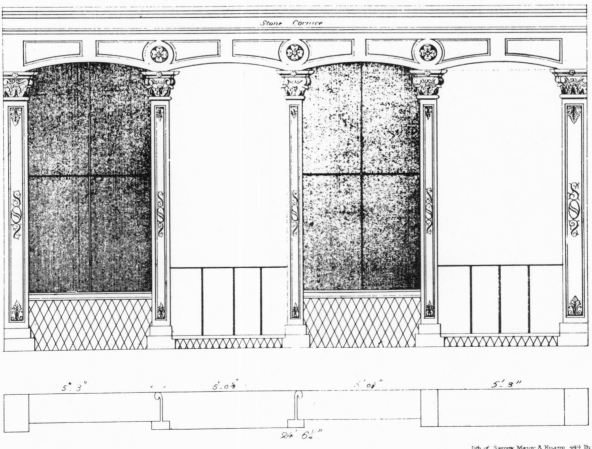

Stone Cornice

ARCHITECTURAL   IRON  WORKS._ NEW YORK

Lith. of Sarony, Major & Knapp. 449 Broadway N.Y.

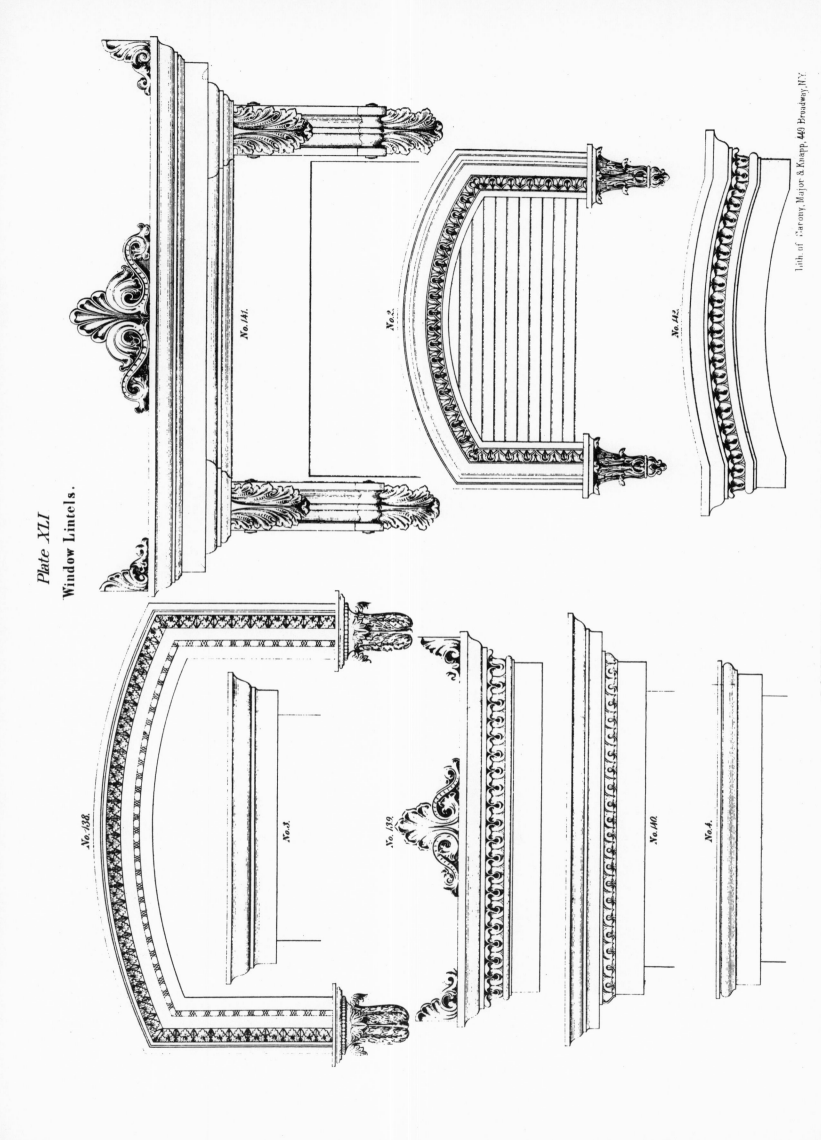

Plate XLI

Window Lintels.

No. 141.

No. 2.

No. 142.

No. 138.

No. 3.

No. 139.

No. 140.

No. 4.

Lith. of Sarony, Major & Knapp, 449 Broadway, N.Y.

ARCHITECTURAL IRON WORKS,—NEW-YORK.

*Plate XLII*

# Window Lintels.

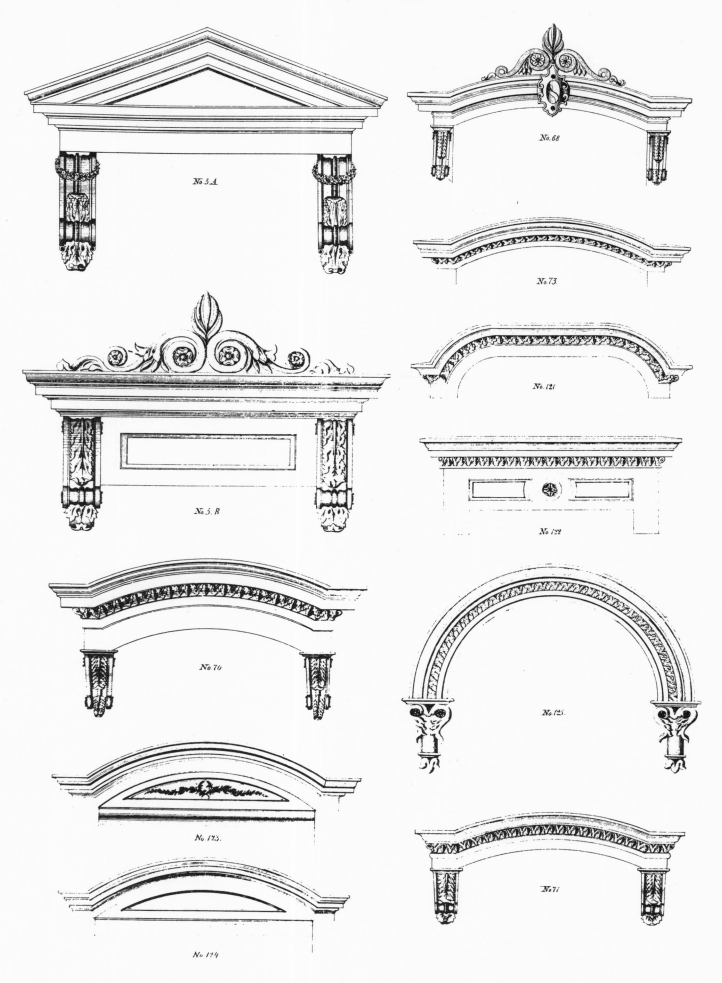

No. 5 A.

No. 68.

No. 73.

No. 5. B

No. 121.

No. 122.

No. 70

No. 125.

No. 123.

No. 71.

No. 124.

Lith. of Sarony Major & Knapp, 449 Broadway

ARCHITECTURAL IRON WORKS NEW YORK

*Plate XLIII*

# Awning Posts & Rod.

Window Lintel Architraves & Sill.

## ARCHITECTURAL IRON WORKS,_NEW-YORK.

Lith of Sarony Major & Knapp 449 B'way, N.Y.

*Plate XLIV*
# Window Lintels. Lamp Post. etc.

No: 145

No: 144

No: 126

No: 104

No: 128

No: 88

Lith of Sarony, Major & Knapp, 449 Broadway, N.Y.

## ARCHITECTURAL IRON WORKS, NEW-YORK

Plate XIV.

Window Lintels, Architraves & Sills.

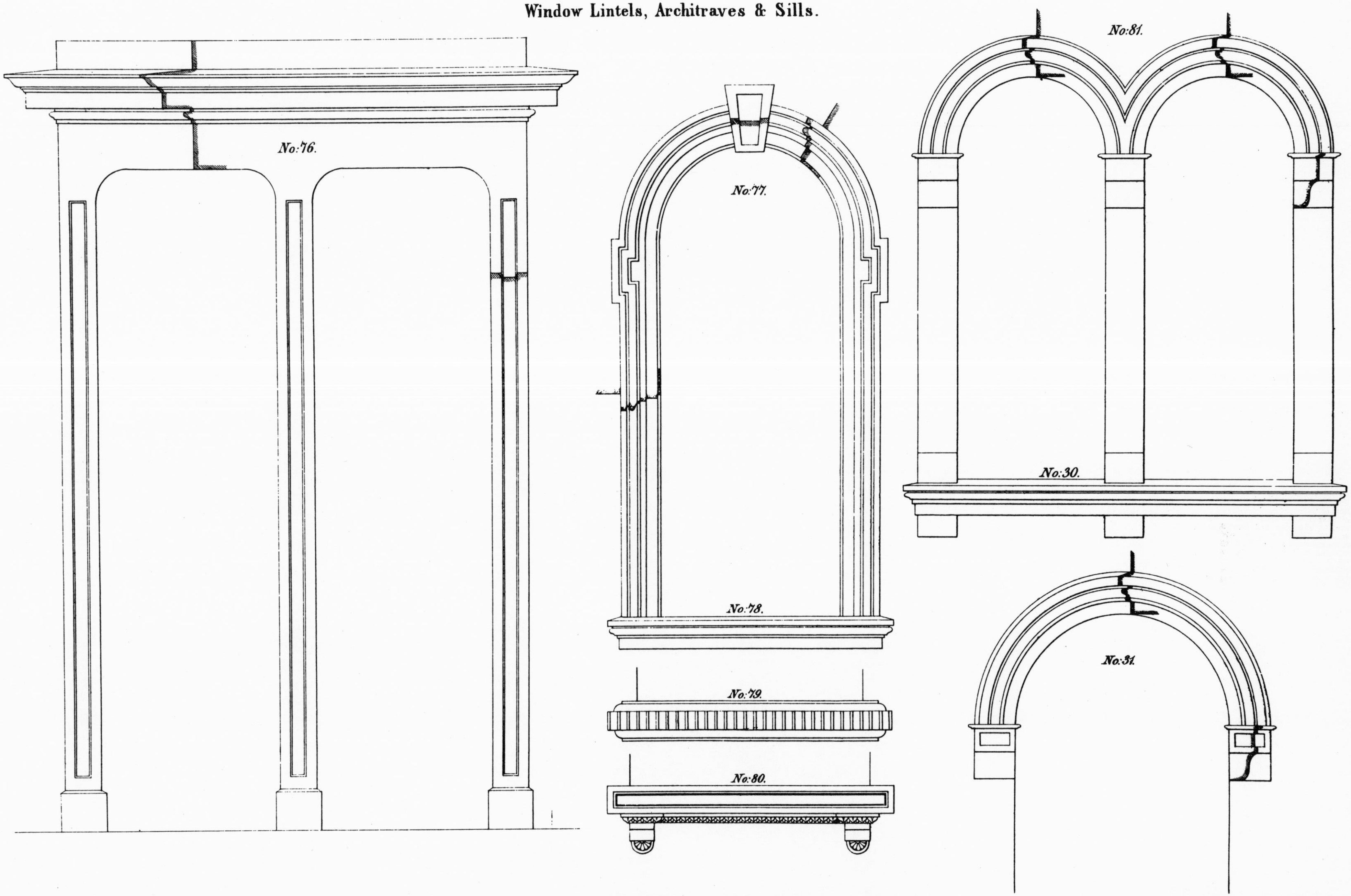

No. 81.

No. 30.

No. 34.

No. 77.

No. 78.

No. 79.

No. 80.

No. 76.

ARCHITECTURAL IRON WORKS,—NEW-YORK.

Lith. of Sarony, Major & Knapp, 449 Broadway, N.Y.

*Plate XLVII*

CLAY BUILDING

ARCHITECTURAL IRON WORKS,—NEW YORK.

*Plate XLVII.*

## Consoles and Brackets.

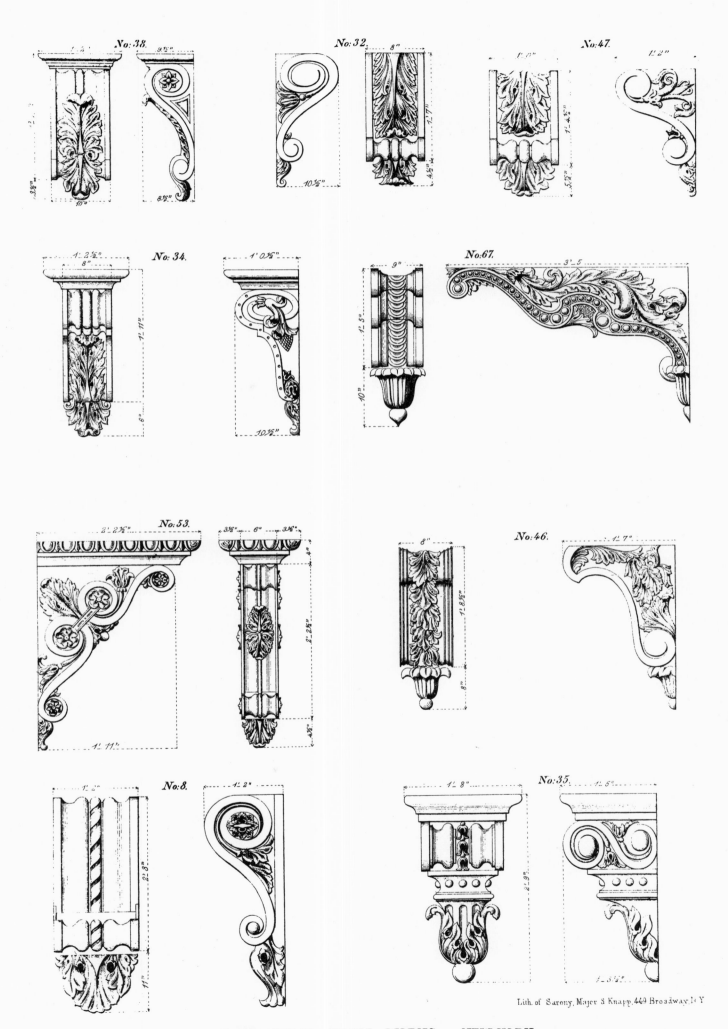

No: 38.   No: 32.   No: 47.

No: 34.   No: 67.

No: 53.   No: 46.

No: 8.   No: 35.

Lith. of Sarony, Major & Knapp, 449 Broadway, N.Y

## ARCHITECTURAL IRON WORKS,_ NEW-YORK

*Plate XLVIII.*
## Consoles Corbels & Urn.

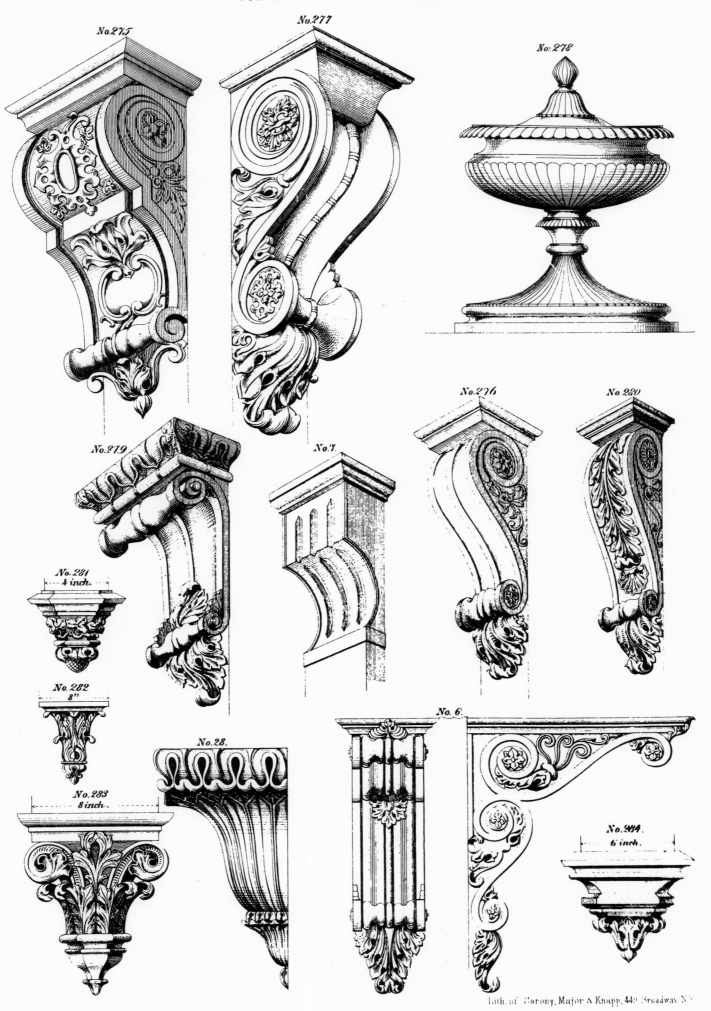

No.275

No.277

No: 278

No.276

No.230

No.279

No.1

No.281
4 inch.

No.282
8"

No.283
8 inch.

No.23.

No. 6.

No.224.
6 inch.

Lith. of Sarony, Major & Knapp, 449 Broadway N.Y.

## ARCHITECTURAL IRON WORKS _ NEW-YORK.

Plate XLIX.

# Elevations and Sections of Columns and Capitals.

## ARCHITECTURAL IRON WORKS,—NEW-YORK.

Plate L.

Gothic Capital.

Stewart Capital.

Stewart Capital.

No 418

No 417

No 416

ARCHITECTURAL · IRON · WORKS, __ NEW-YORK.

Lith of Sarony Major & Knapp. 449 Broadway N.Y.

Plate LI.

No. 157.

No. 158.

No. 159.

No. 160.

No. 161.

Corinthian Order.

Composite Order.

Ionic Order.

Doric Order.

Tuscan Order.

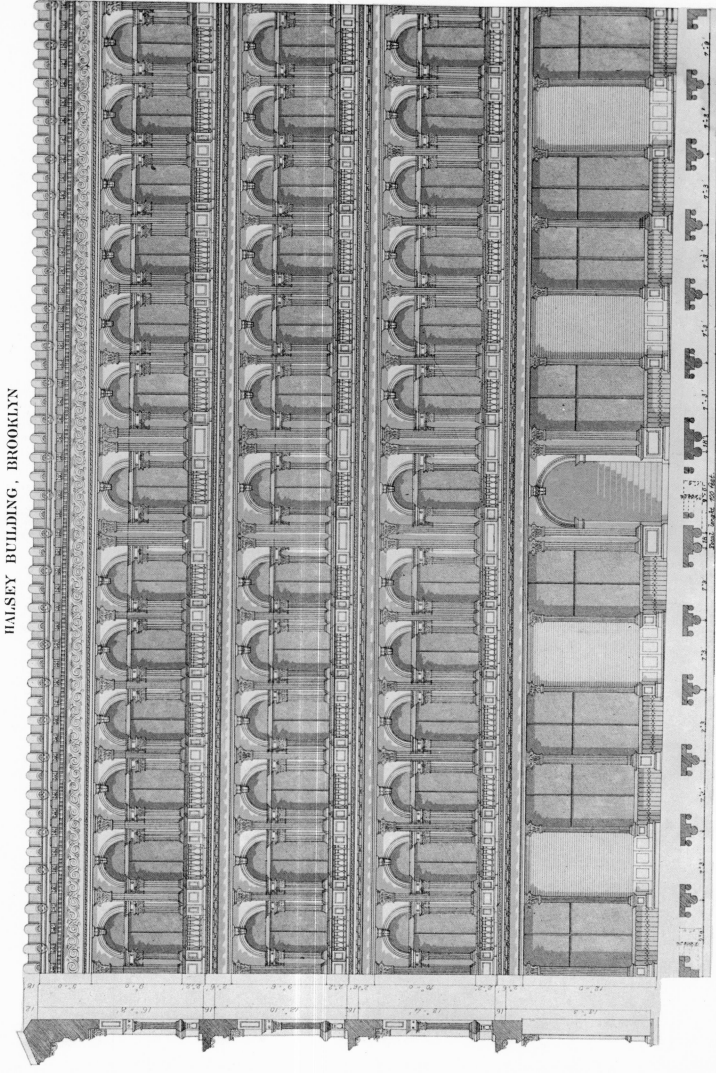

*Plate LIl.*
*No. 13.*

HALSEY BUILDING, BROOKLYN

ARCHITECTURAL IRON WORKS – NEWYORK.

Litz & Barry, Mayr & Knapp, 115 Broadway, N.Y.

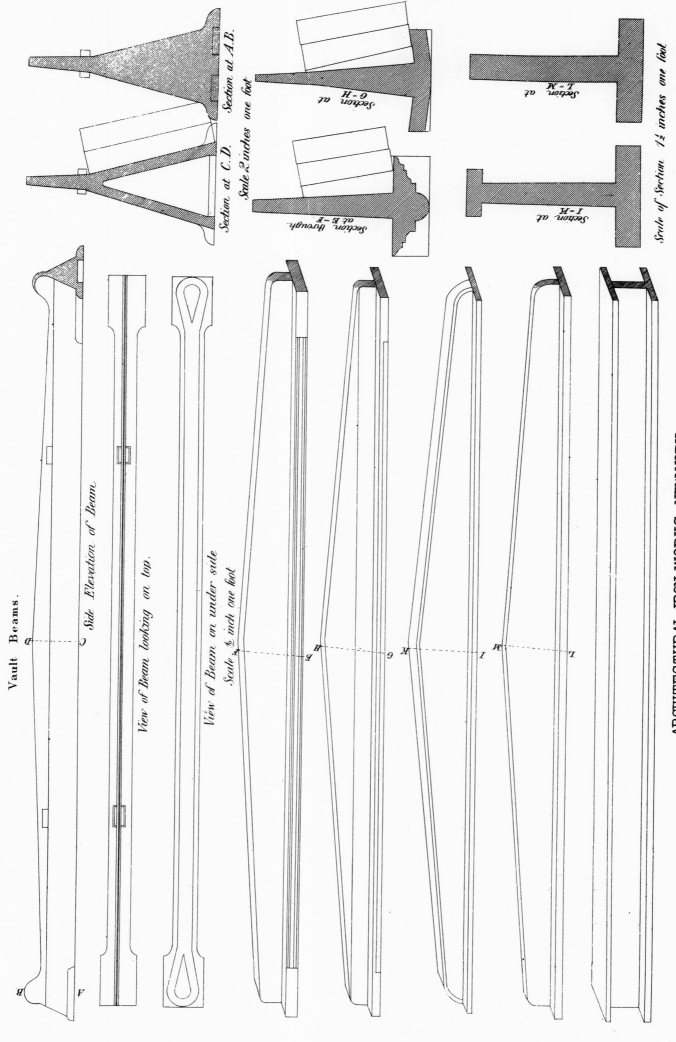

Plate LIII.

Vault Beams.

Side Elevation of Beam

View of Beam looking on top.

View of Beam on under side

Scale ½ inch one foot

Section at C.D.    Section at A.B.

Scale 2 inches one foot

Section through at E−F.    Section at G−H.

Section at T−M.    Section at I−H.

Scale of Section ½ inches one foot

Lith. d. Sarony Major & Knapp, 449 Broadway N.Y.

ARCHITECTURAL IRON WORKS — NEW YORK.

*Plate LIV.*

*No.3.*

ARCHITECTURAL IRON WORKS,—NEW-YORK.

*Plate LV.*

Design for Front of Dwelling House.

*No 41*

Lith. of Sarony Major & Knapp. 449 Broadway. N.Y.

ARCHITECTURAL IRON WORKS,—NEW-YORK.

# PATENT METALLIC WINDOW BLINDS
## BURGLAR & FIRE-PROOF.

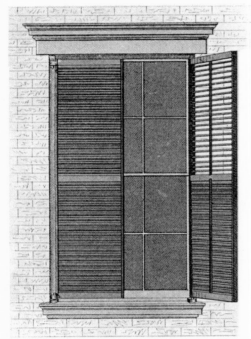

ELEVATION showing Blinds for OUTSIDE USE

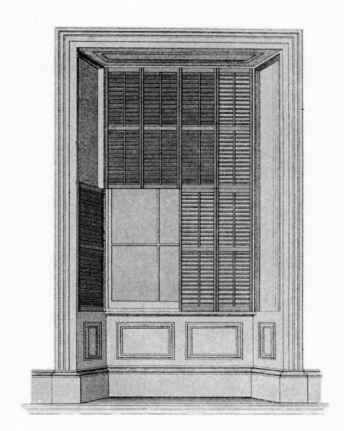

ELEVATION showing Blinds for INSIDE USE.

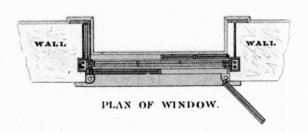

PLAN OF WINDOW.

This Blind obviates all the difficulties and inconveniences of the wooden Blind, and is designed to supersede the folding iron shutter and the outside and inside wooden blind and shutter. It is fire-proof and by actual experiment is shown to resist the fire much longer than the ordinary iron shutter; and water thrown on it while hot will not curve, warp or open it so as to expose the window to the flames. This Blind does not shrink, warp or settle by exposure to solar or artificial heat or by atmospheric changes, thus freeing it from those objections to the wooden blind, which so try the patience of House keepers. The wires are always in order and can not be pulled out, the slats remain unbroken and can be so adjusted as to let in the exact amount of light and air required. It is self fastening and fastenings are always in order. It is substantial and, unlike the wooden blind, requires little or no repairs and is capable of the highest finish and ornament. The inside blinds are specially adapted to first class dwellings, churches &c.

Many of the first Architects and Builders of this and other Cities have given these Blinds their unqualified approval. They have been adopted by various banking houses and dwellings and recently by the new Court House in Brooklyn.

Manufactured for the American Iron Blind Company and orders received for the same by the Architectural Iron Works, 42 Duane Street, New York.

f Sarony, Major & Knapp, 449 Broadway, N.Y.

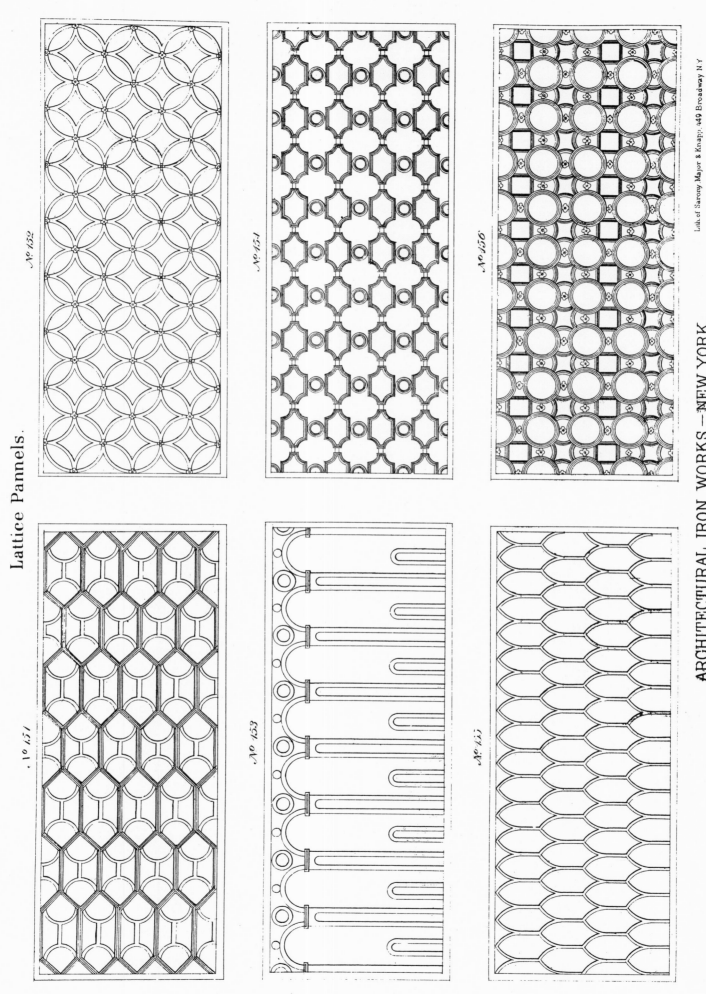

*Plate LVII.*

Lattice Pannels.

N.º 152

N.º 154

N.º 156

N.º 151

N.º 153

N.º 155

ARCHITECTURAL IRON WORKS — NEW YORK.

Lith. of Sarony Major & Knapp. 449 Broadway N.Y.

*Plate LVIII*

JOHN C. GRAY ESQ. BOSTON.

*No. 11*

ARCHITECTURAL IRON WORKS, NEW-YORK

Lith. of Sarony, Major & Knapp, 449 Broadway. N.Y.

Plate LIX.

Elevation of H.H. Honnewell's Building, Boston.
No. 12.

ARCHITECTURAL IRON WORKS,— NEW-YORK

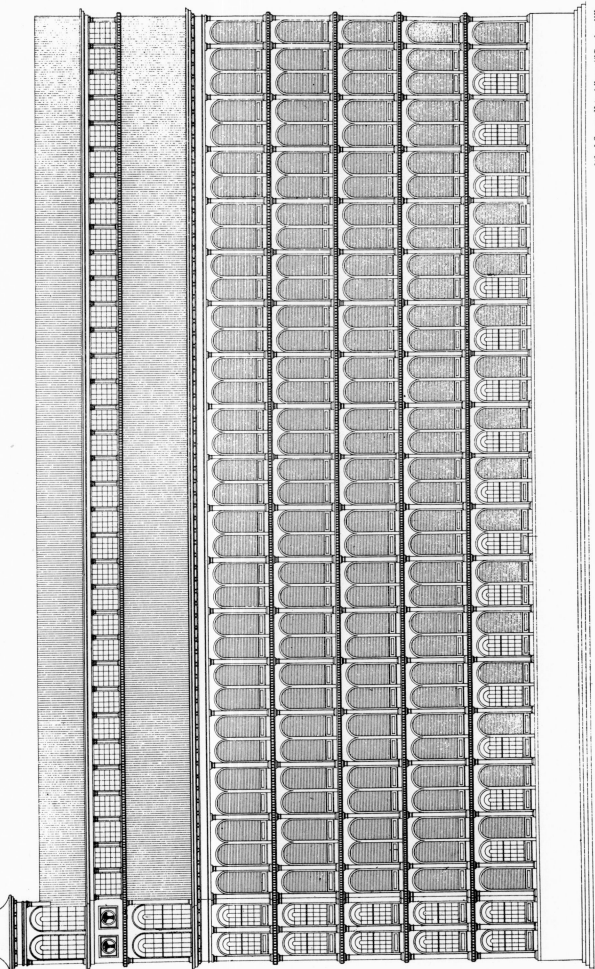

*Plate IX.*

**Elevation of Grain Building.**

*N°. 27.*

ARCHITECTURAL IRON WORKS,—NEW-YORK.

Lith. of Sarony, Major & Knapp, 449 Broadway, N.Y.

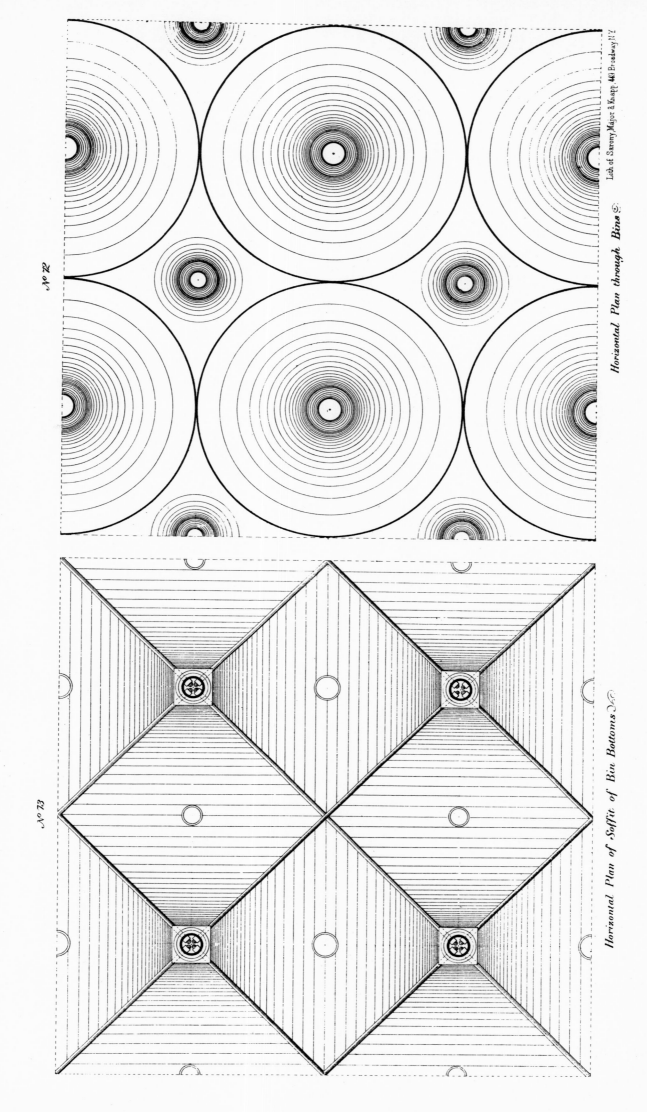

*Plate LXI.*

Details of Grain Building.

*N? 72*

*N? 73*

*Horizontal Plan through Bins C.*

*Horizontal Plan of Soffit of Bin Bottoms N°C.*

Lith of Sarony Major & Knapp. 449 Broadway N.Y.

ARCHITECTURAL IRON WORKS,— NEW-YORK.

*Plate LXII.*
Section of Grain Building through Bins.
*Nº 71*

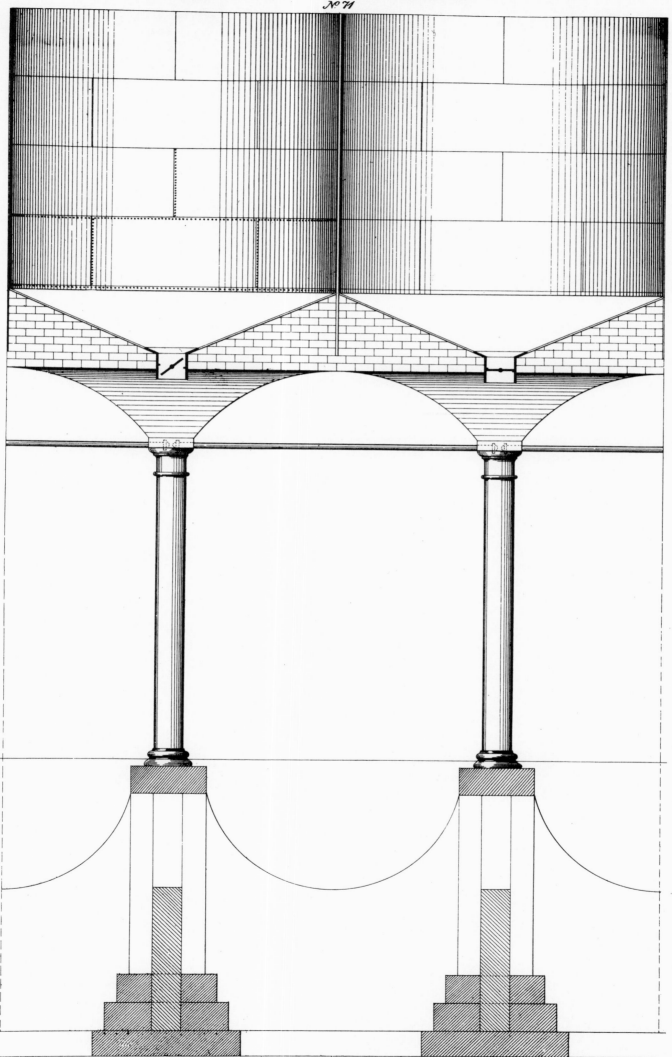

ARCHITECTURAL IRON WORKS ⎯ NEW-YORK.

Plate LXIII

Tension Rod Girders.    No 271

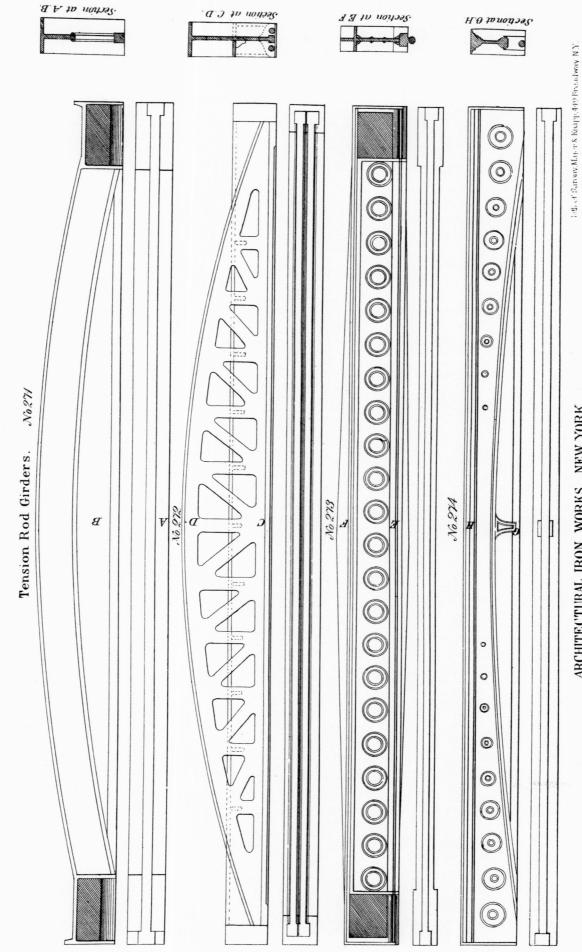

Section at A.B.

Section at C.D.

Section at E.F.

Section at G.H.

No 272

No 273

No 274

Lith. of Sarony Major & Knapp. 449 Broadway N.Y.

ARCHITECTURAL IRON WORKS.  NEW YORK.

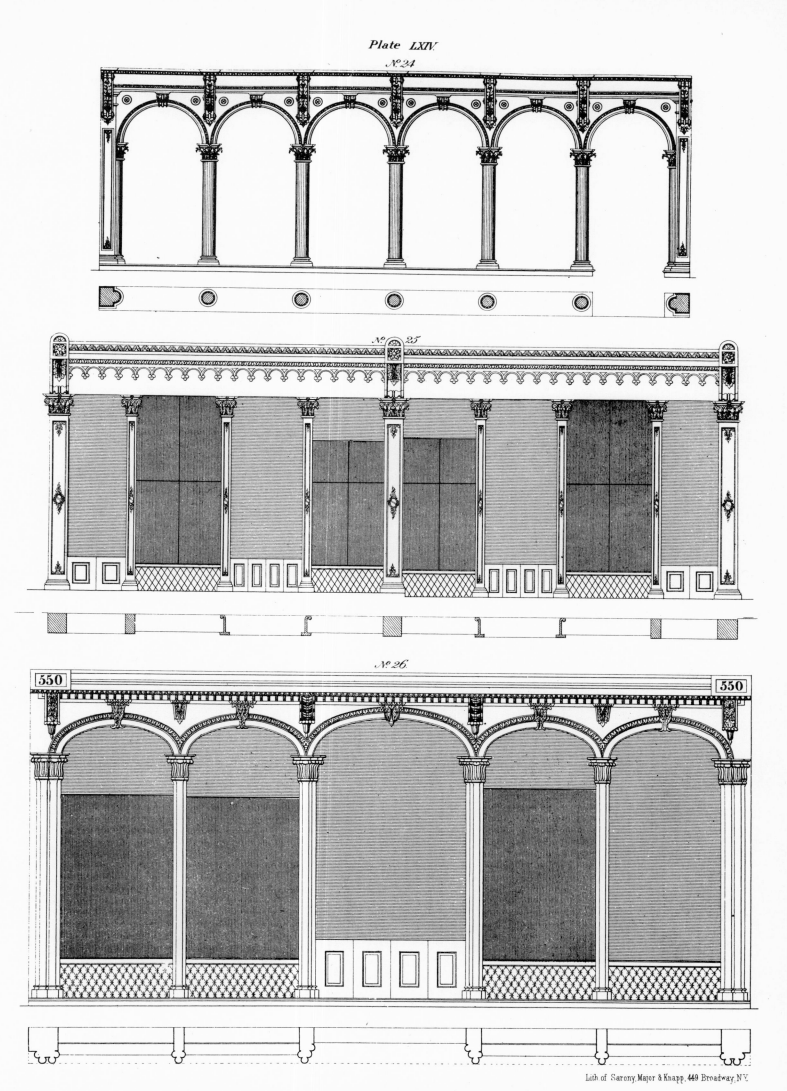

Plate LXIV.
No. 24

No. 25

No. 26
550  550

ARCHITECTURAL IRON WORKS,—NEW-YORK.

Lith of Sarony, Major & Knapp, 449 Broadway, N.Y.

*Plate LXV.*
*No. 23.*

ARCHITECTURAL IRON WORKS,—NEW-YORK.

Lith of Sarony, Major & Knapp, 449 Broadway N.Y.

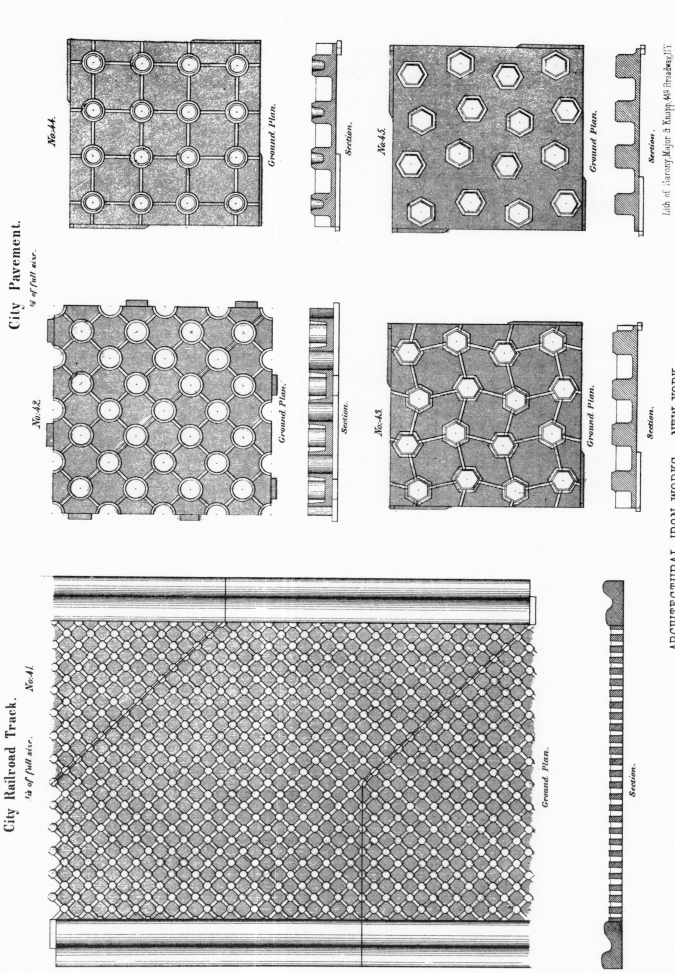

Plate LXVI.

City Pavement.
⅙ of full size.

No. 44.

Ground Plan.

Section.

No. 45.

Ground Plan.

Section.

No. 42.

Ground Plan.

Section.

No. 43.

Ground Plan.

Section.

City Railroad Track.
⅛ of full size.

No. 41.

Ground Plan.

Section.

Lith. of Sarony, Major & Knapp, 449 Broadway, N.Y.

ARCHITECTURAL IRON WORKS,— NEW-YORK.

*Plate LXVII.*
## Elevation for Banking House & Office.
*No: 22*

## ARCHITECTURAL IRON WORKS NEWYORK

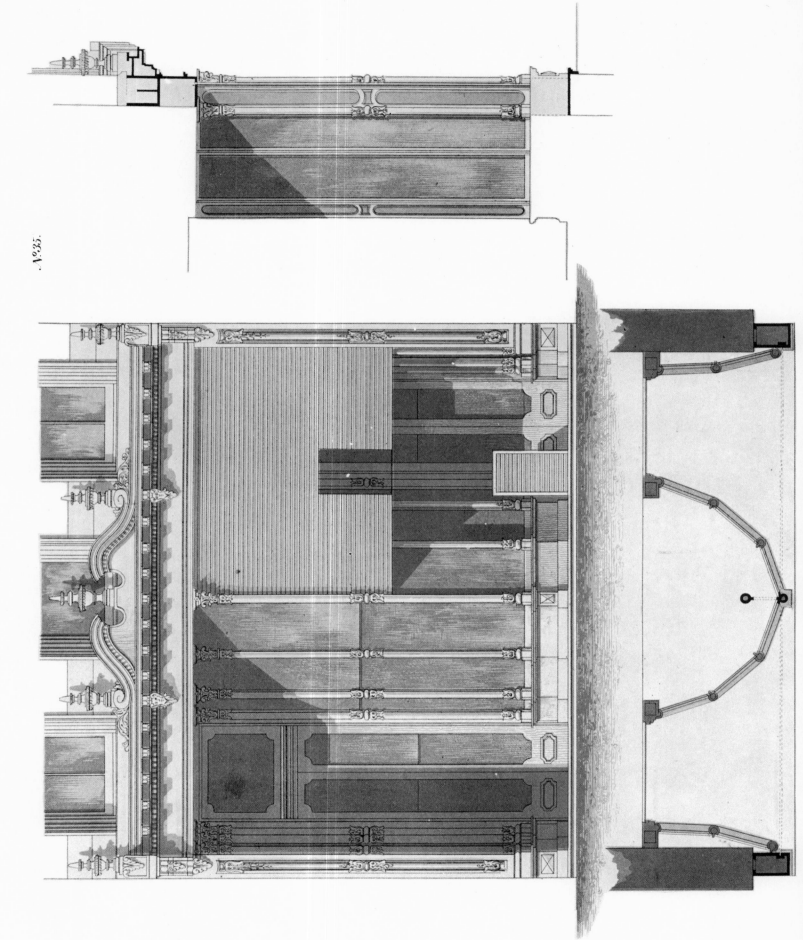

Plate LXVIII.

ARCHITECTURAL IRON WORKS — NEW YORK.

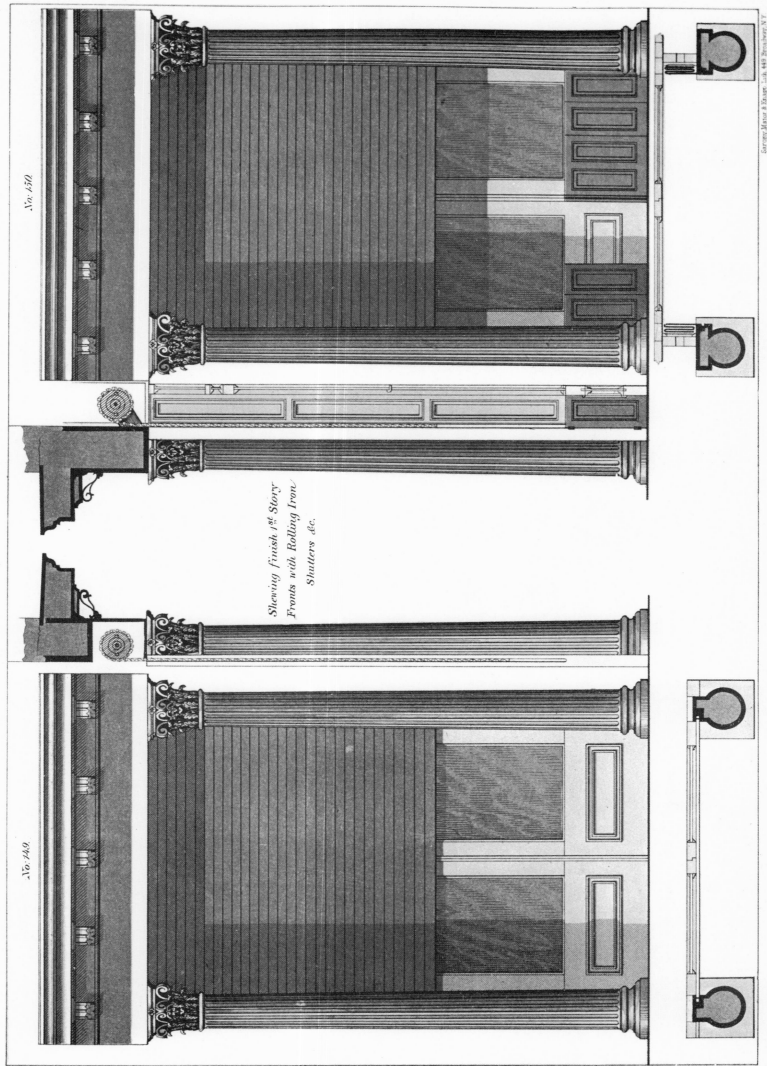

Plate LXIX.

No: 150.

No: 149.

Shewing Finish 1st Story
Fronts with Rolling Iron
Shutters &c.

Sarony Major & Knapp. Lith 449 Broadway N.Y.

ARCHITECTURAL IRON WORKS NEW YORK

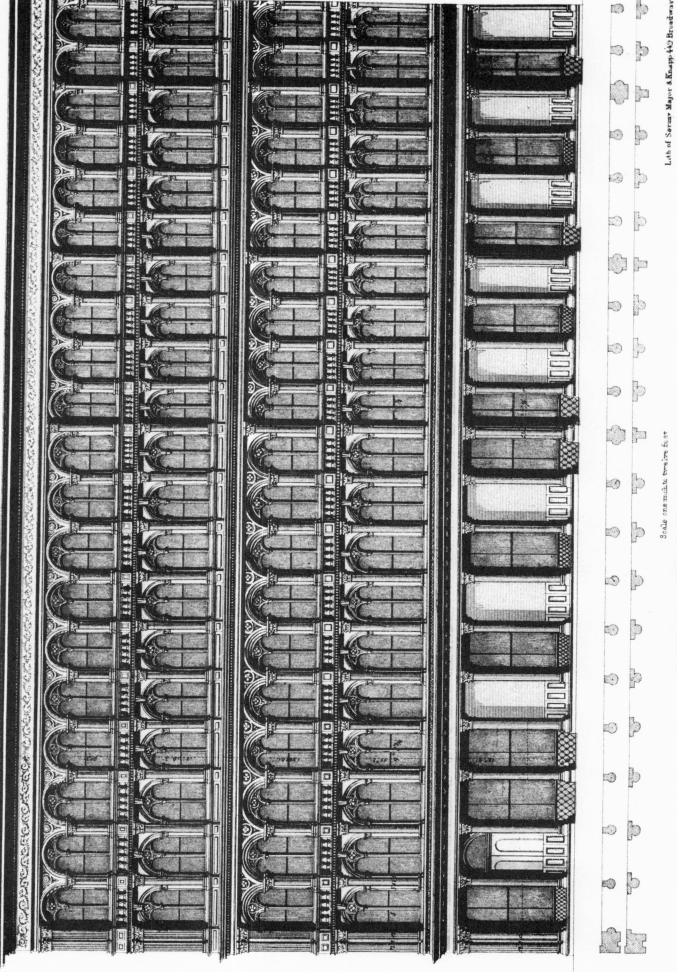

Scale one and a twelve in feet

ARCHITECTURAL IRON WORKS NEW YORK

Plate LXVI.

Details: Rolling Iron Shutters.

Section of Shutter; ⅓ full size.

E

No. 5.

D

B

O

A

ARCHITECTURAL IRON WORKS — NEW-YORK.

# Consoles Brackets & Rosett's.

No.33.

No.36.

No.37.

No.39.

No.40.

No. 13.

No. 12.

No. 11.

No. 10.

No. 16.

No. 14.

No. 9.

No. 15.

No.56.
9"

No.66.
7 ½"

No.69.
6"

No.54.

No. 55.
12"

3' 0"

No.57.
6"

No. 72.
7½"

No.65.
7½"

No.59.
4"

No.60.
10"

No.63.
4"

No.64.
3¾"

No.74.
4½"

No.75.
3½"

## ARCHITECTURAL IRON WORKS, — NEW-YORK.

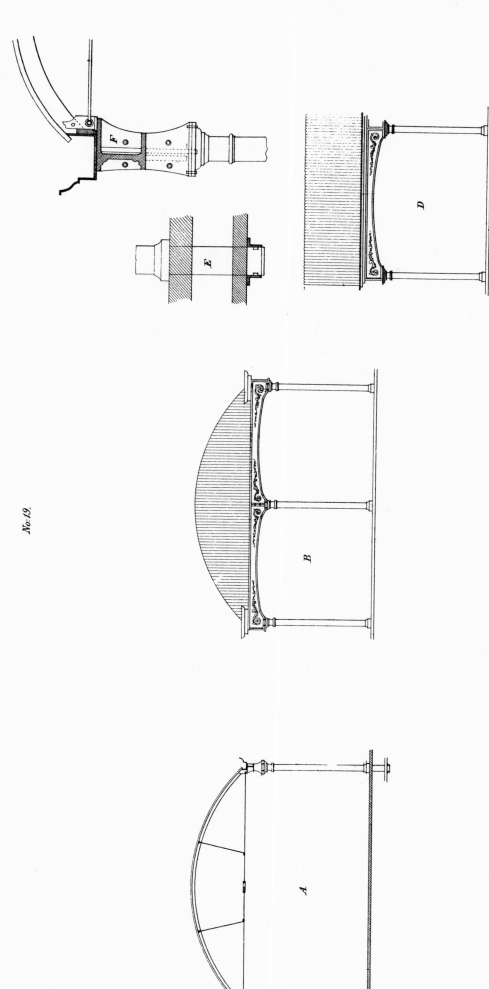

*Plate LXXIII*

## Sugar Shed for Havana, Cuba.

*No.19.*

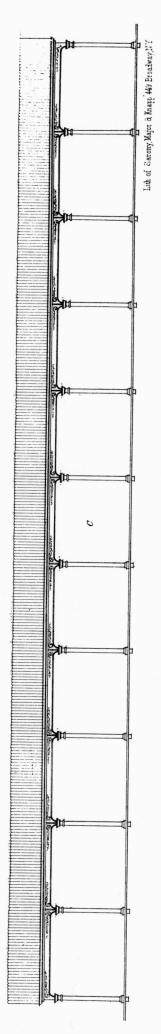

ARCHITECTURAL IRON WORKS,—NEW-YORK

*Plate LXVII*

N°36.

BILLING & Co.

ARCHITECTURAL IRON WORKS — NEW-YORK.

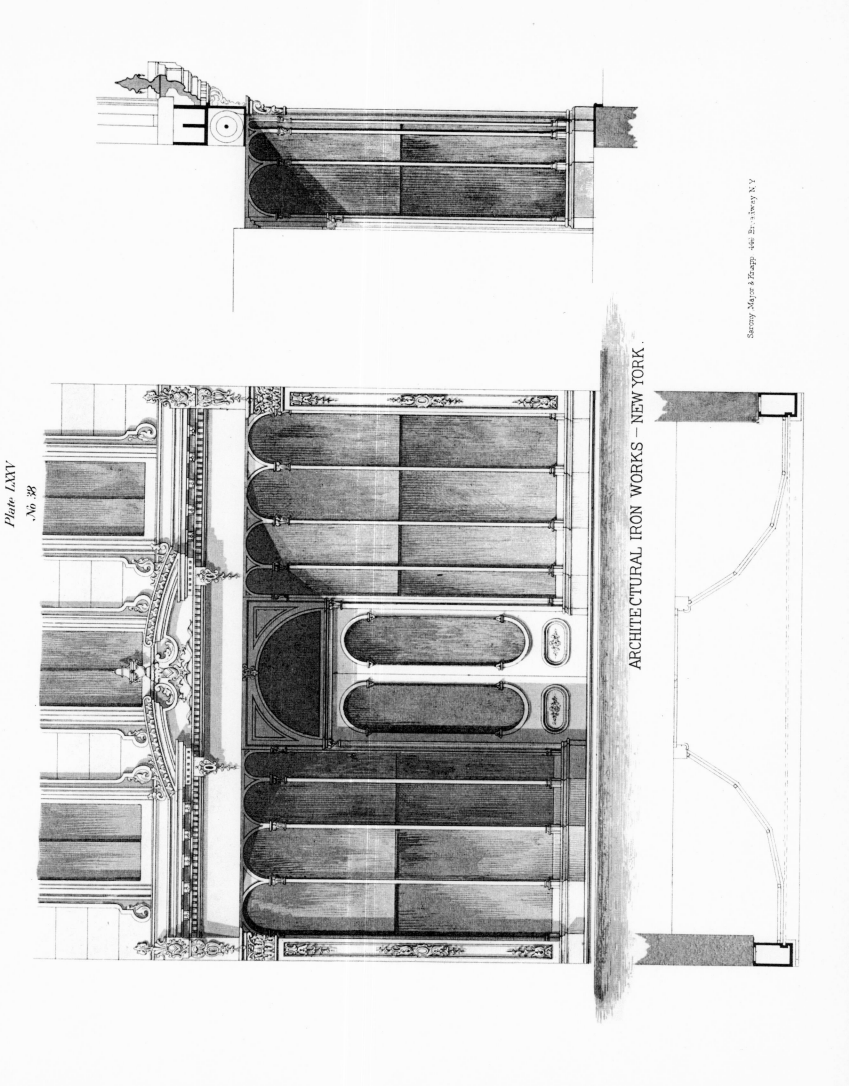

Plate LXV

Nº 38

ARCHITECTURAL IRON WORKS — NEW YORK.

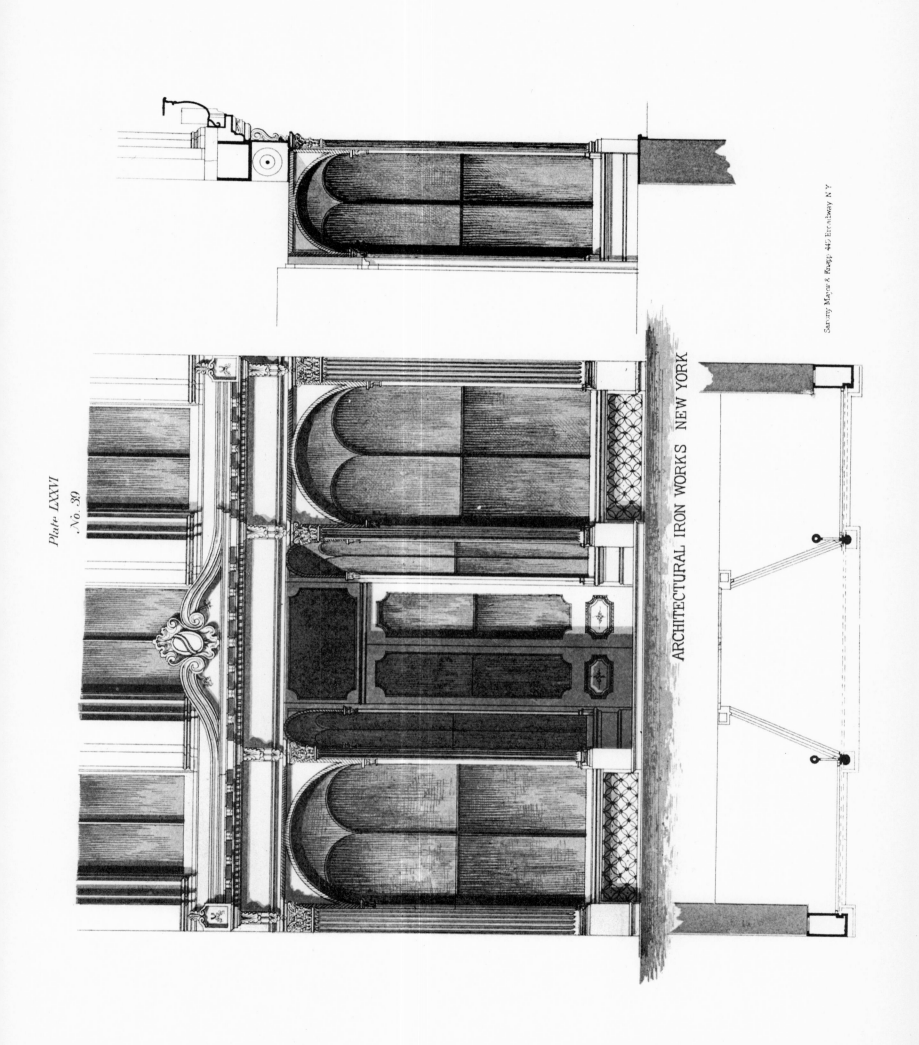

*Plate LXXVI*

*No. 80*

ARCHITECTURAL IRON WORKS  NEW YORK

Sarony, Major & Knapp, 449 Broadway N.Y.

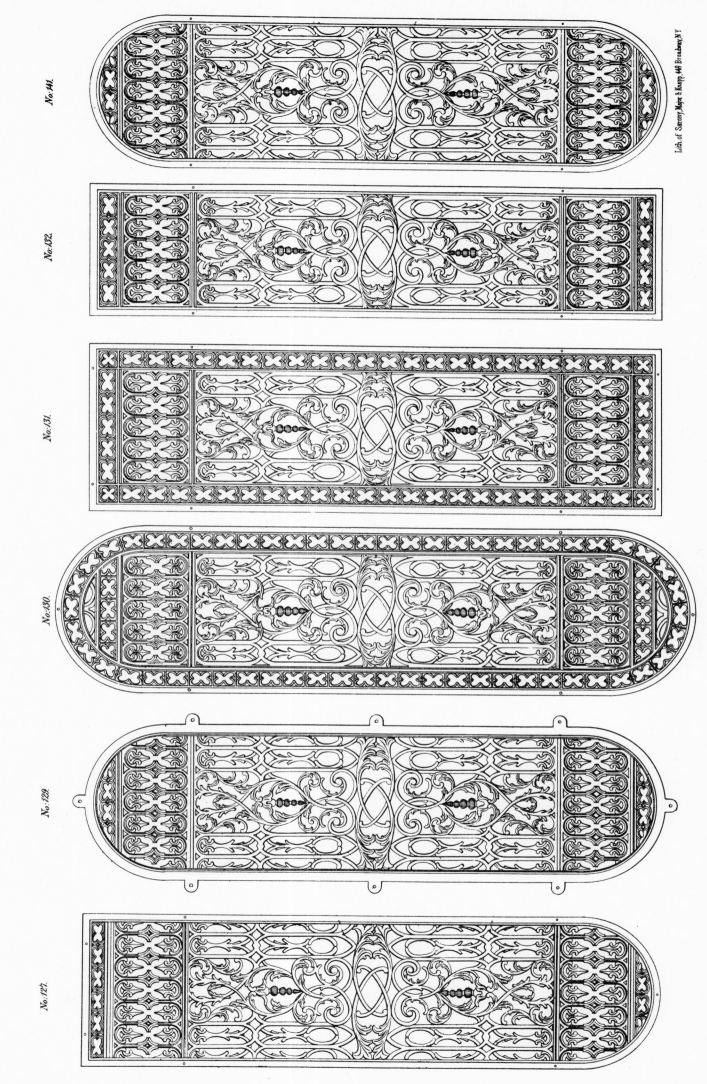

Plate LXXVII

No. 141.

No. 132.

No. 131.

No. 130.

No. 129.

No. 127.

ARCHITECTURAL IRON WORKS.—NEW-YORK.

Lith. of Sarony, Major & Knapp, 449 Broadway, N.Y.

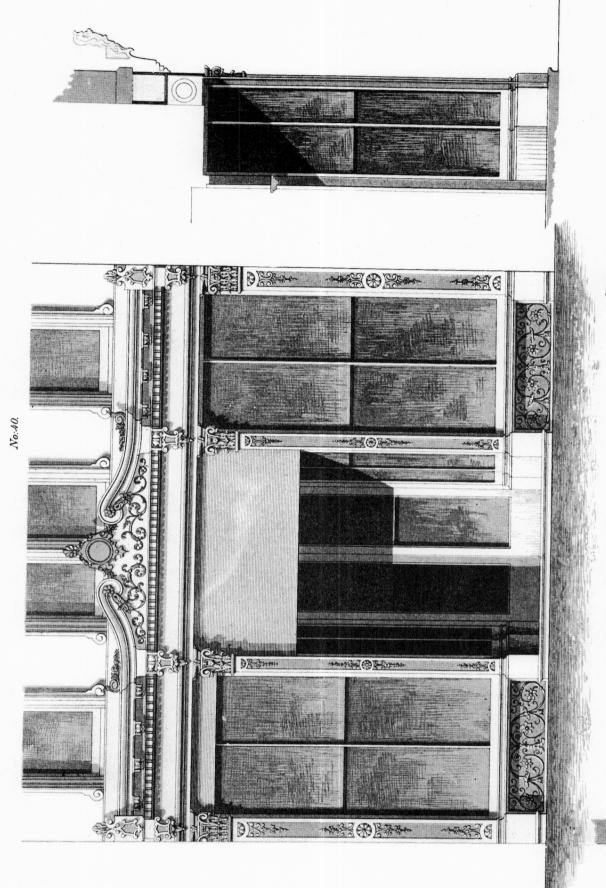

Plate LXVIII

No. 40.

ARCHITECTURAL IRON WORKS,—NEW-YORK.

Lith. of Sarony Major & Knapp 449 Broadway N.Y

*Plate LXXX*

ARCHITECTURAL IRON WORKS __ NEW-YORK.

Lith of Sarony Major & Knapp, 449 Broadway, N.Y

Plate LXXX

No: 163.

No: 165.

No: 166.

No: 164.

1' 1"

5½"

3' 10"

3' 8"

2' 3"

1' 4"

11"

1' 7"

No: 167.

No: 168.

No: 169.

2' 2"

1' 8"

2' 4"

9"

8½"

1' 2"

No: 170.

No: 171.

No: 172.

3' 0"

4' 0"

2' 9"

10"

1' 6"

8"

Lith. of Sarony, Major & Knapp, 449 Broadway, N.Y.

ARCHITECTURAL IRON WORKS,—NEW-YORK.

Plate LXXVI.

No: 196.

No: 175.

No: 183.

No: 184.

No: 182.

No: 181.

No: 180.

No: 179.

No: 178.

No: 177.

No: 124.

No: 173.

Lith of Sarony Major & Knapp. 44 Broadway N.Y.

ARCHITECTURAL IRON WORKS,—NEW-YORK.

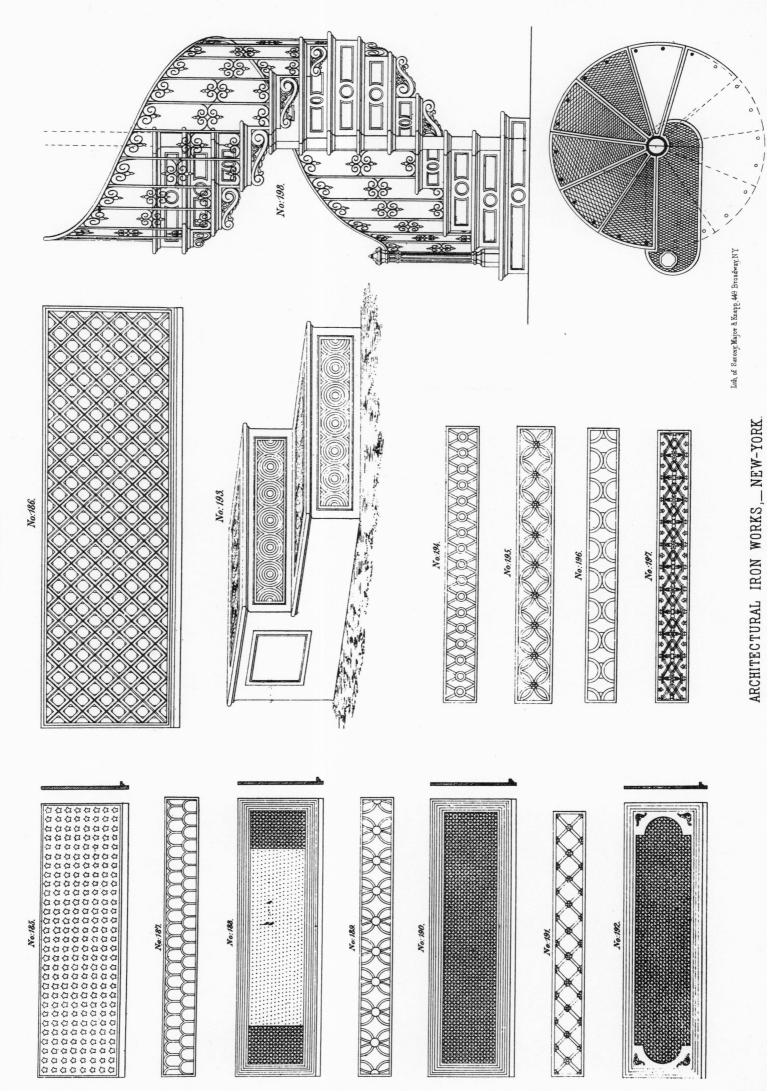

Plate LXXXII

No: 198.

No: 186.

No: 193.

No: 194.

No: 195.

No: 196.

No: 197.

No: 185.

No: 187.

No: 188.

No: 189.

No: 190.

No: 191.

No: 192.

Lith. of Sarony, Major & Knapp, 449 Broadway, N.Y.

ARCHITECTURAL IRON WORKS,—NEW-YORK.

*Plate LXXVIII.*

Elevation & Section of Sidewalk &c. Shewing Vault under Street.

*No. 45.*

Lith of Sarony, Major & Knapp 449 Broadway N.Y.

ARCHITECTURAL IRON WORKS,— NEW-YORK.

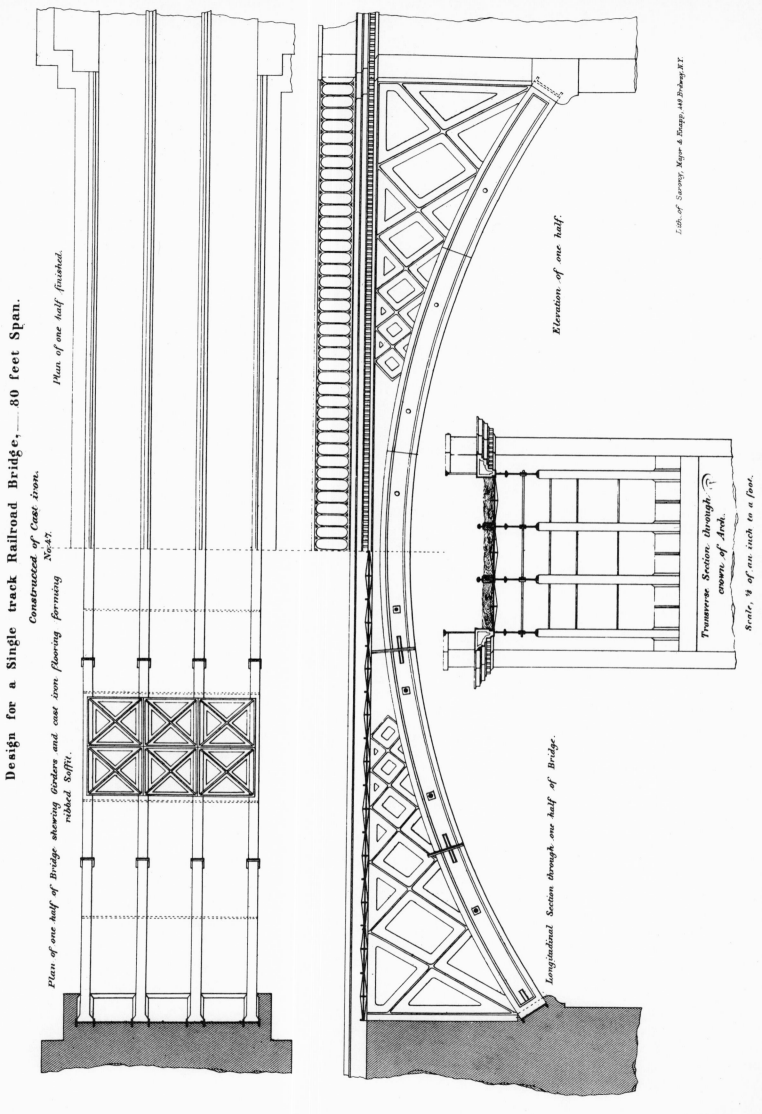

*Plate LXXXIV*

Design for a Single track Railroad Bridge,⸺ 80 feet Span.

*Constructed of Cast iron,*
*No: 47.*

*Plan of one half finished.*

*Plan of one half of Bridge shewing Girders and cast iron flooring forming*
*ribbed Soffit.*

*Longitudinal Section through one half of Bridge.*

*Elevation of one half.*

*Transverse Section through*
*crown of Arch.*

*Scale, ¼ of an inch to a foot.*

*Lith. of Sarony, Major & Knapp, 449 Broadway, N.Y.*

ARCHITECTURAL IRON WORKS, NEW-YORK.

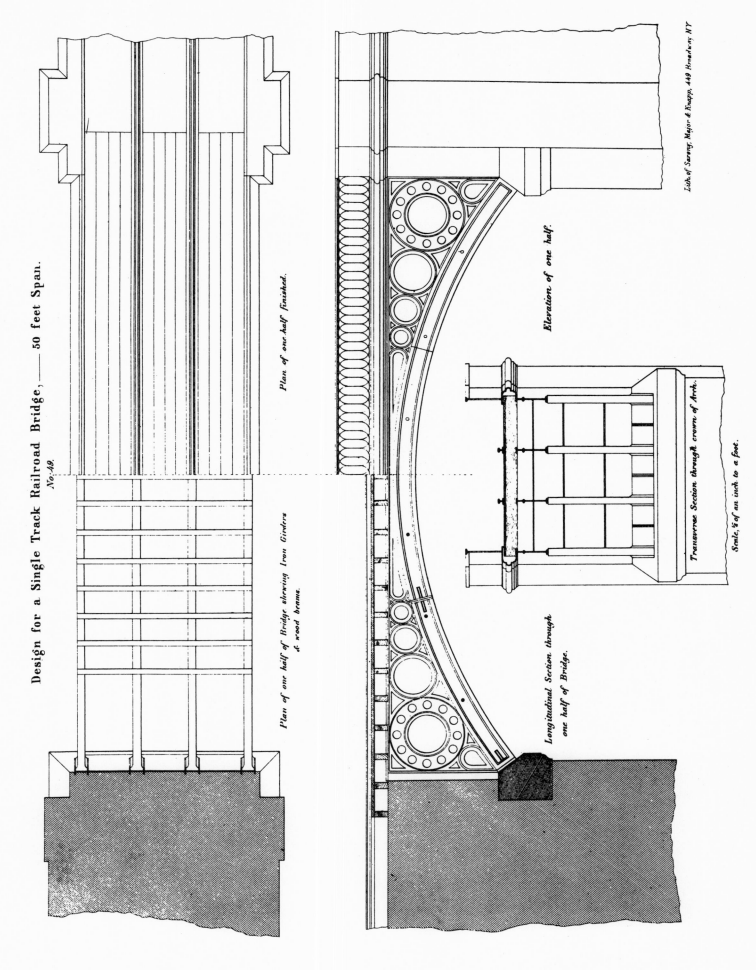

Plate LXXV

Design for a Single Track Railroad Bridge, ___ 50 feet Span.
No. 49.

Plan of one half finished.

Plan of one half of Bridge shewing Iron Girders
& wood beams.

Elevation of one half.

Transverse Section through crown of Arch.

Scale, ¼ of an inch to a foot.

Longitudinal Section through
one half of Bridge.

Lith. of Sarony, Major & Knapp, 449 Broadway, N.Y.

ARCHITECTURAL IRON WORKS, ___ NEW-YORK.

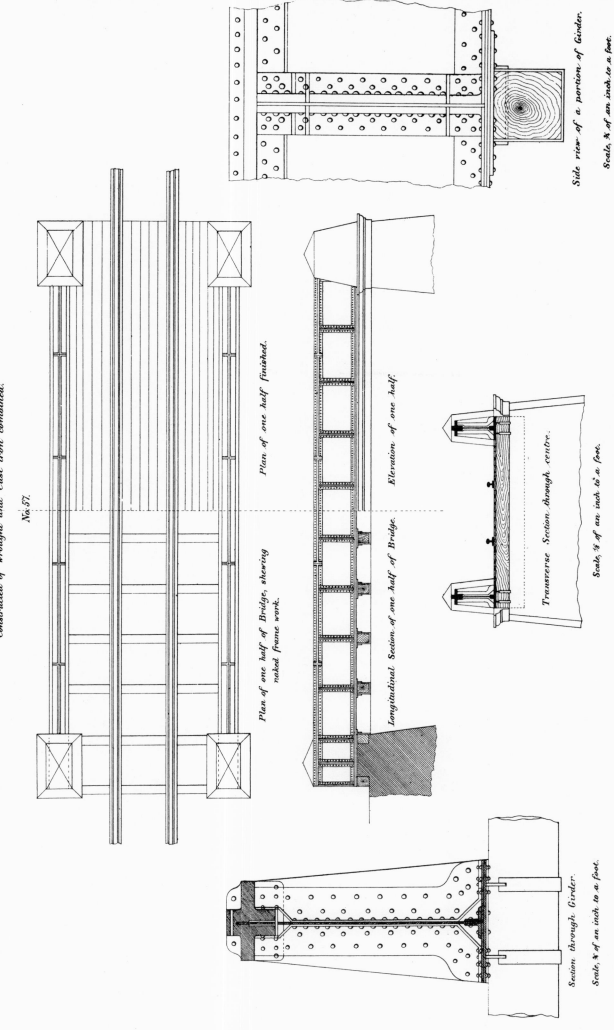

*Plate LXXVI*

## Design for a Single track Railroad Bridge, 40 feet between bearings.

*Constructed of wrought and Cast iron combined.*

*No. 57.*

Plan of one half finished.

Plan of one half of Bridge, shewing
naked frame work.

Elevation of one half.

Longitudinal Section of one half of Bridge.

Side view of a portion of Girder.

*Scale, ¾ of an inch to a foot.*

*Lith. of Sarony, Major & Knapp, 449 Broadway, N.Y.*

Transverse Section through centre.

*Scale, ¾ of an inch to a foot.*

Section through Girder.

*Scale, ¾ of an inch to a foot.*

ARCHITECTURAL IRON WORKS,—NEW-YORK.

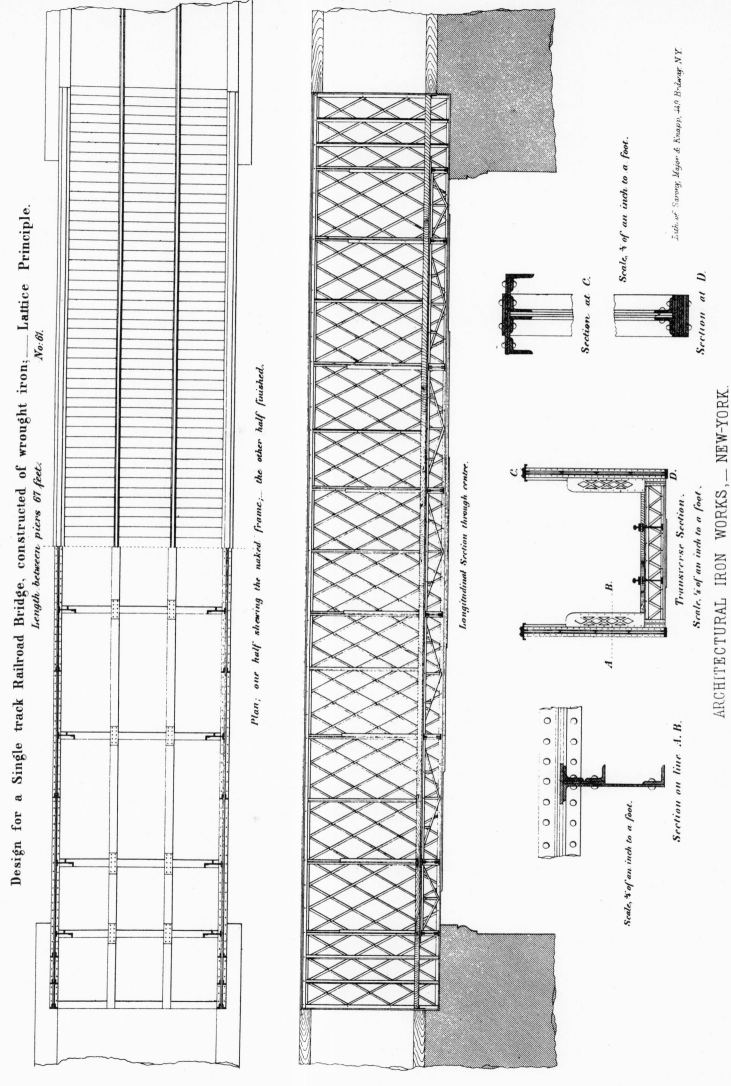

*Plate LXXVII*

**Design for a Single track Railroad Bridge, constructed of wrought iron; — Lattice Principle.**

*No: 61.*

*Length between piers 67 feet.*

*Plan; one half shewing the naked frame; — the other half finished.*

*Longitudinal Section through centre.*

*Section at C.*

*Scale, ⅜ of an inch to a foot.*

*Section at D.*

*Lith.ᵈ Sarony, Major & Knapp, 449 B—way N.Y.*

*Section on line A. B.*

*Scale, ⅜ of an inch to a foot.*

*Transverse Section.*

*Scale, ½ of an inch to a foot.*

**ARCHITECTURAL IRON WORKS, — NEW-YORK.**

*Plate LXXXVIII*

Ferry House, Built for Dr. Thomas Rainey, Rio Janeiro, Brazil.

*No. 74.*

FERRY.

BOTAFOGO E CATTETE

NICTHEROHY

ARCHITECTURAL IRON WORKS.— NEW-YORK.

*Plate LXXXIX*

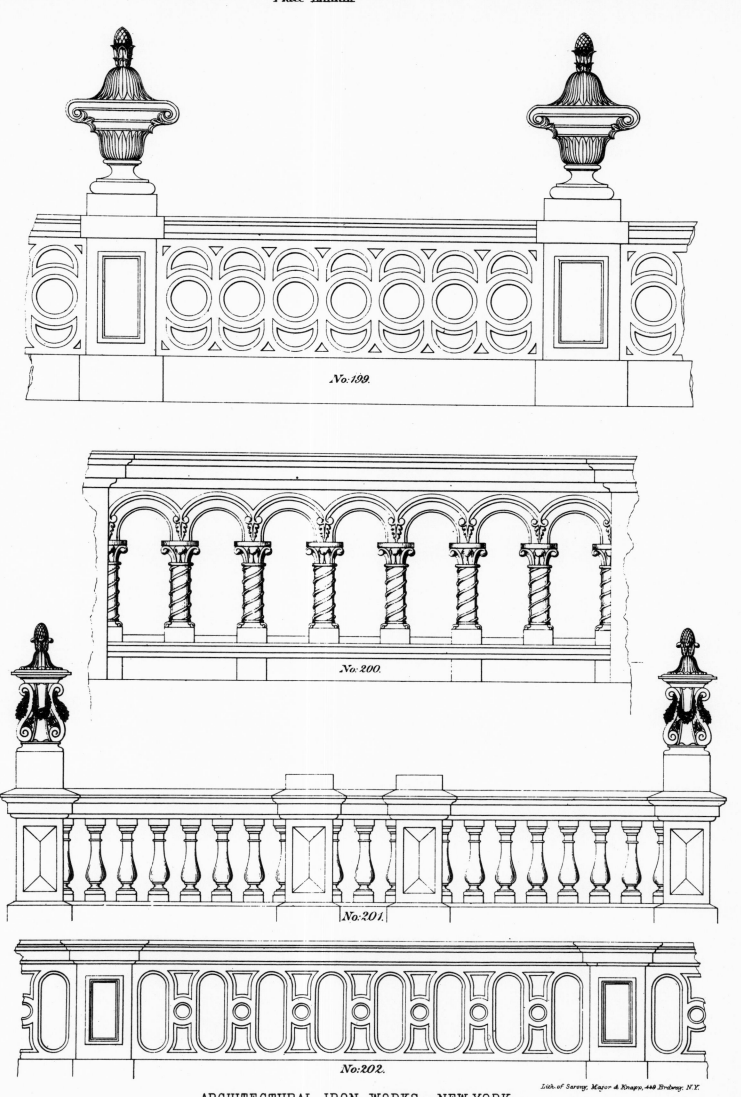

*No: 199.*

*No: 200.*

*No: 201.*

*No: 202.*

ARCHITECTURAL IRON WORKS,—NEW-YORK.

*Plate XC*

*No: 14.*

ARCHITECTURAL IRON WORKS, __ NEW-YORK.

Lith of Sarony, Major & Knapp. 449 Broadway N.Y.

*Plate XCI*

*No: 203.*

*No: 48.*

*No: 204.*

*No: 205.*

*No: 206.*

*No: 207.*

ARCHITECTURAL IRON WORKS,___NEW-YORK.

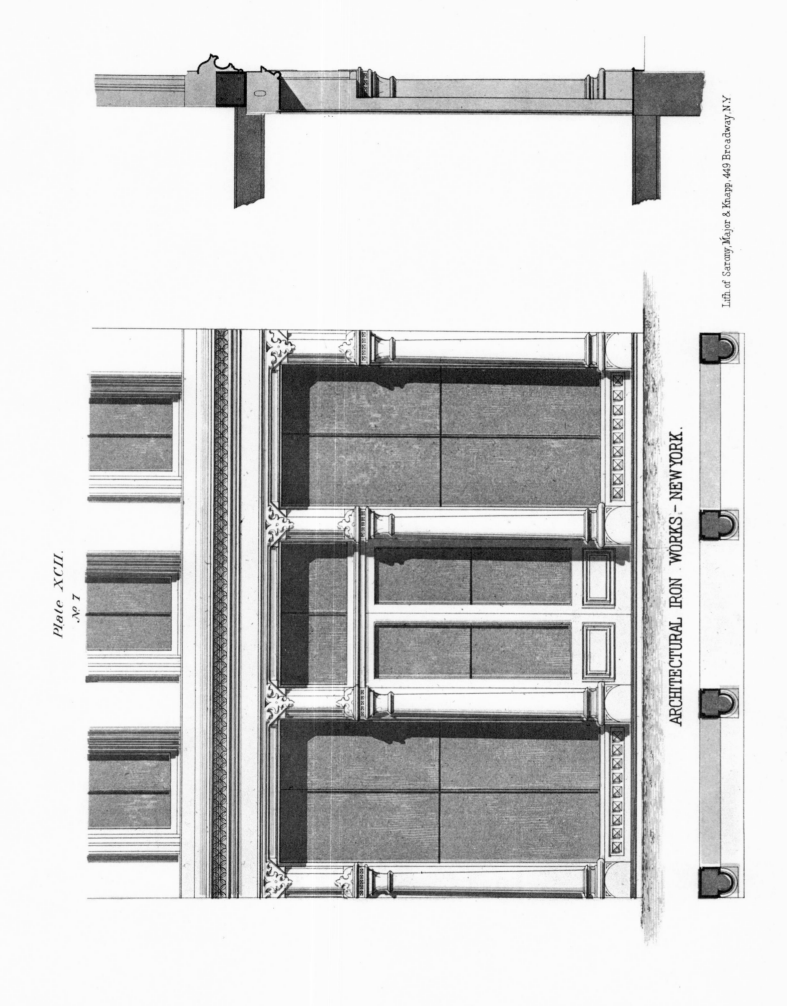

Plate XCII.

№ 7

ARCHITECTURAL IRON WORKS – NEWYORK.

Lith. of Sarony, Major & Knapp, 449 Broadway, N.Y.

*Plate XCIII.*

No 208

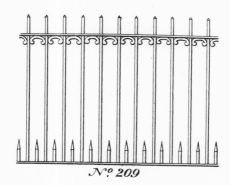

No 209

No 210

No 211

No 212

No 213

No 214

No 215

No 216

No 217

No 218

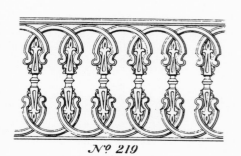

No 219

ARCHITECTURAL IRON WORKS,—NEW-YORK.

Lith. of Sarony Major & Knapp, 449 Bd'w'y N.Y.

*Plate XCIV*

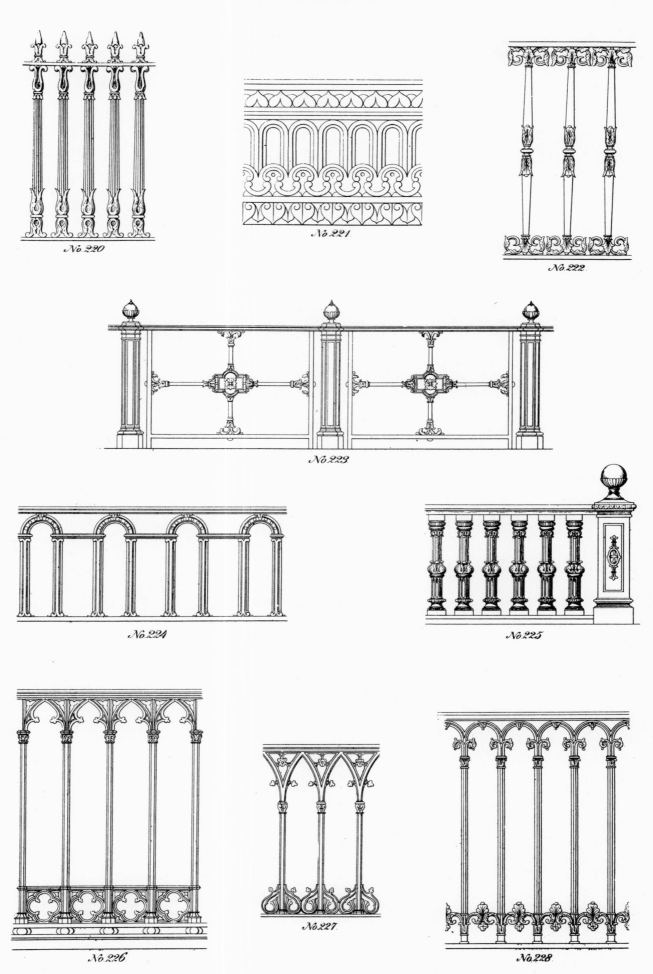

No. 220

No. 221

No. 222

No. 223

No. 224

No. 225

No. 226

No. 227

No. 228

ARCHITECTURAL IRON WORKS — NEW YORK.

Lith. of Sarony, Major & Knapp, 449 Broadway, N.Y.

*Plate XCV*

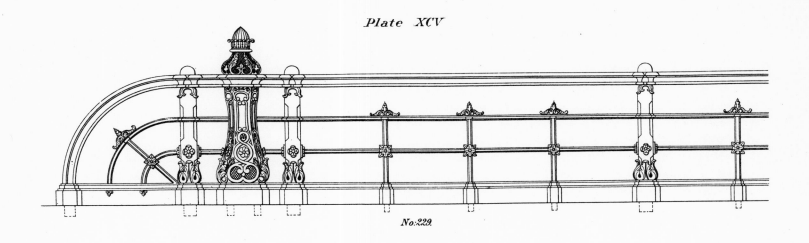

No: 229.

No: 230.

No: 231.

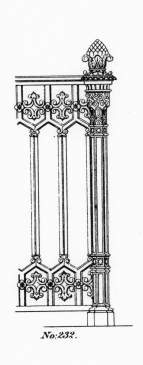

No: 232.

No: 233.

No: 234.

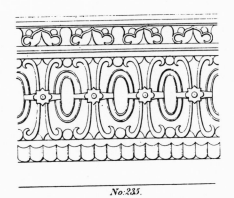

No: 235.

Lith. of Sarony, Major & Knapp, 449 Broadway, N.Y.

ARCHITECTURAL IRON WORKS,___ NEW-YORK.

*Plate XCVI.*

No: 236.

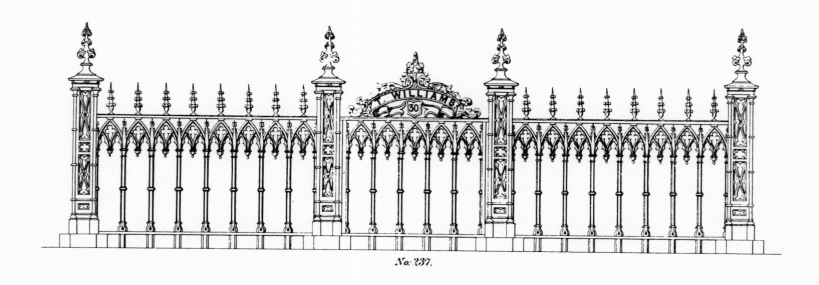

No: 237.

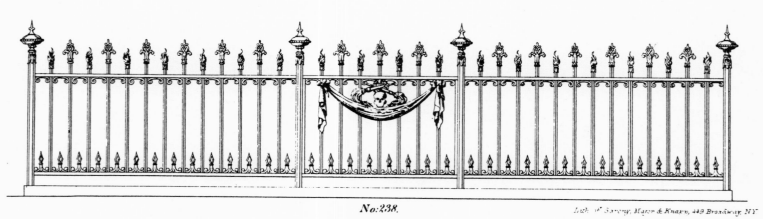

No: 238.

ARCHITECTURAL IRON WORKS, — NEW-YORK.

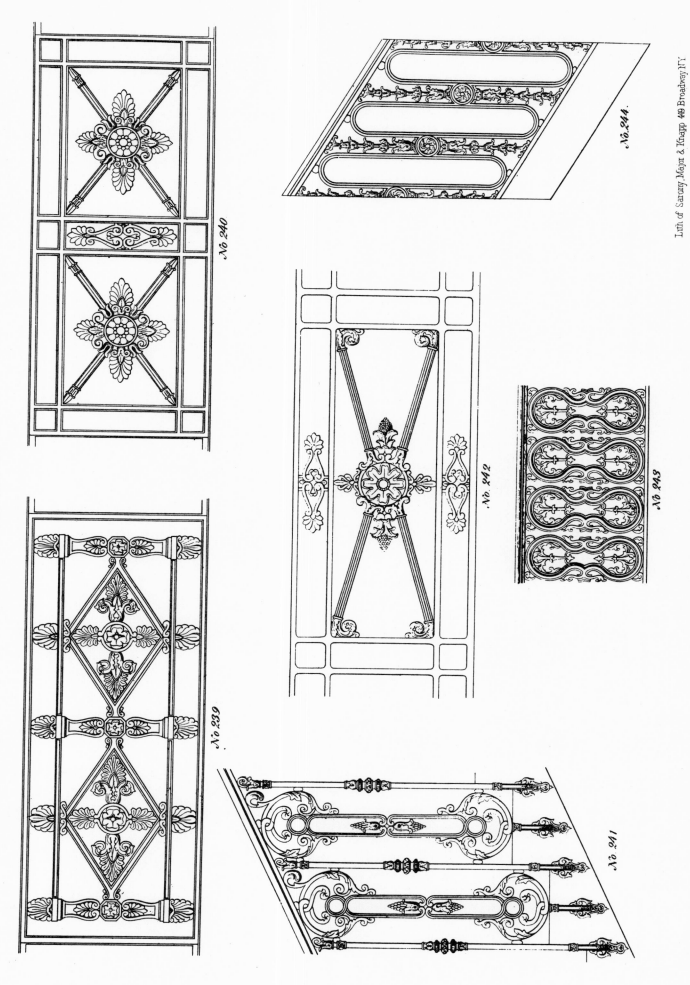

Plate XCVII

Nô 240

Nô 239

Nô 242

Nô 243

Nô 244.

Nô 241

Lith of Sarony, Major & Knapp 449 Broadway N.Y.

ARCHITECTURAL IRON WORKS.- NEW YORK

Plate XCVIII

No. 246.

No. 247.

No. 248.

No. 249.

No. 250.

No. 251.

No. 252.

No. 253.

No. 254.

No. 255.

No. 256.

Lith of Sarony, Major & Knapp, 449 Brdway N Y.

ARCHITECTURAL IRON WORKS,——NEW-YORK.

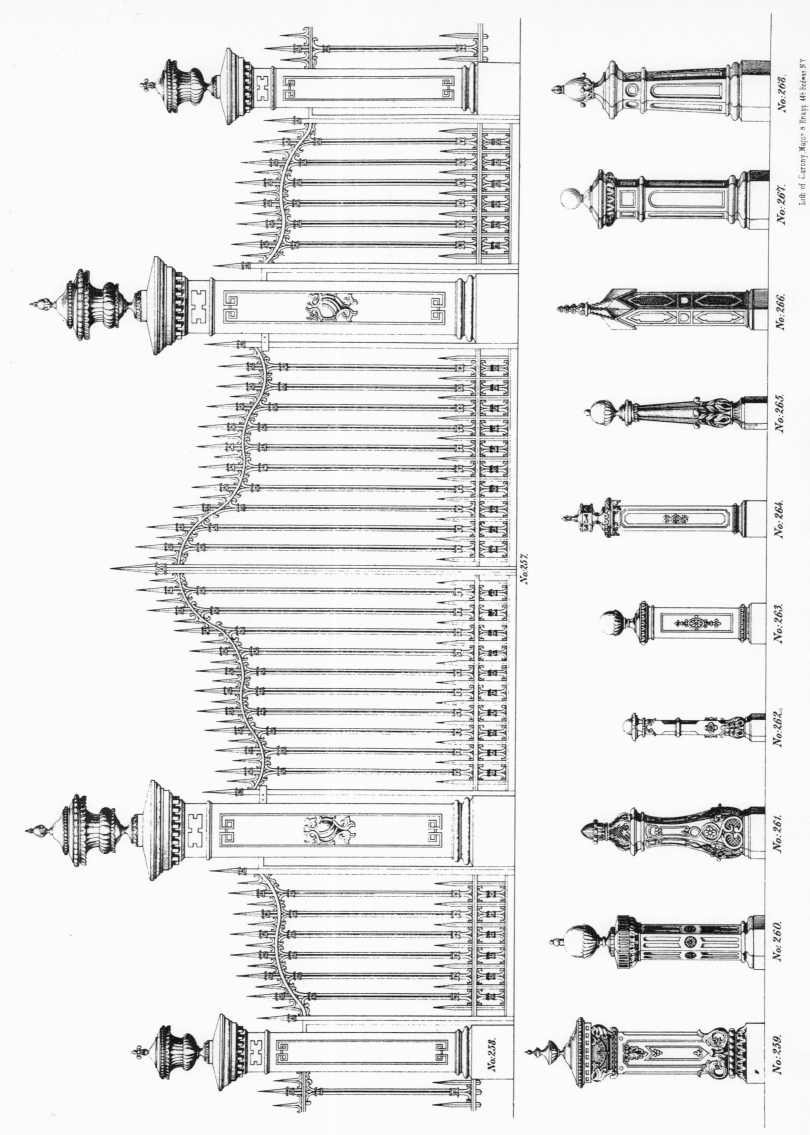

Plate XCIX.

No. 268.

No. 267.

No. 266.

No. 265.

No. 264.

No. 263.

No. 262.

No. 261.

No. 260.

No. 259.

No. 257.

No. 258.

ARCHITECTURAL IRON WORKS,——NEW-YORK.

Lith. of Sarony Major & Knapp 449 Broadway N.Y

*Plate C.*

*No: 105.*

Lith of Sarony, Major & Knapp, 449 Brdway, N.Y.

# ARCHITECTURAL IRON WORKS,——NEW-YORK.

*Plate CI.*

*No: 162.*

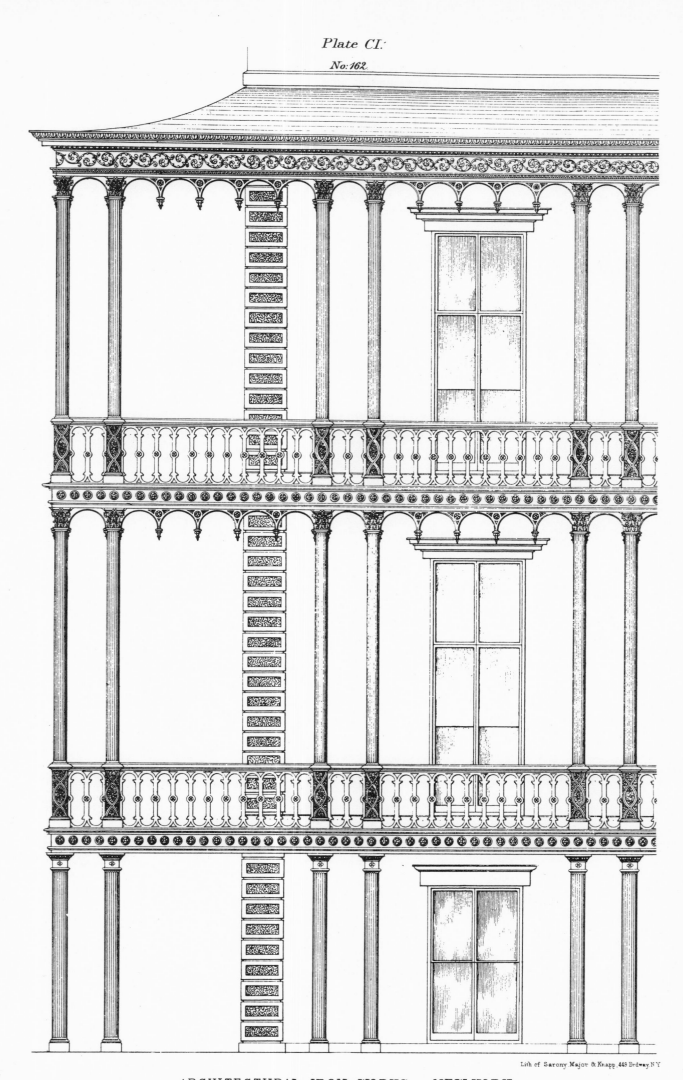

Lith of Sarony Major & Knapp, 449 Brdway, N.Y.

ARCHITECTURAL IRON WORKS,__ NEW-YORK.

*Plate CII.*

*Nº 75.*

ARCHITECTURAL IRON WORKS, — NEW-YORK.

# CAST IRON BUILDINGS:
## Their Construction and Advantages

By James Bogardus

# INTRODUCTION

It is not the purpose of the remarks which follow to pass judgment on the claims of a controversial pamphlet written over a hundred years ago but rather to place it in proper historical perspective.

James Bogardus was born in Catskill, New York, in 1800, and died in New York City in 1874. Apprenticed at the age of fourteen to a watchmaker, he early revealed an inventive turn of mind. With the exception of his patent for iron buildings, his principal inventions were in the field of precision instruments. Among these were engraving and die-cutting machines, a pyrometer, a deep sea sounding device, a machine for pressing glass, and a chronometer. In 1839, in London, he won an award for the best engraving machine and plan for making postage stamps. It is possible that at this time he saw Fairbairn's prefabricated iron building, which was exhibited in London before being shipped to Constantinople.

By 1840 Bogardus was back in New York. In the years that followed he set to work devising his own system of iron construction. After considerable production difficulties, his five-story factory building at the corner of Center and Duane Streets was completed in 1850. Said to have been the first *complete* iron building, it had already been preceded by several structures, also by Bogardus, having cast iron fronts only. One of these, at Washington and Murray Streets in downtown New York, is still standing.* Its interior walls are of brick and its floors of timber.

Unlike his rival and near neighbor, Daniel Badger, Bogardus did not fabricate his own buildings. First and foremost the scientist and inventor, he sublet the actual work of casting—in the case of one building, to six different foundries. *Cast Iron Buildings* is about the system of construction that he devised, and related matters. Scientifically interesting for a number of reasons, the document is far from clear in other respects. Its real interest for us today would seem to lie more in the questions it raises than in those that it answers.

First, there is the question of authorship. Why did Bogardus allow his name to appear on the title page when, on the next, he goes out of his way to tell us that the real author is John W. Thomson? Is it the voice of Thomson or Bogardus that is speaking? Second, there is the claim that Bogardus is the inventor of "the first complete cast iron edifice ever erected in America, or in the world"—where is the evidence to support this claim? Finally, what of the patent? The pamphlet gives a detailed description of how the elements of the *front* are assembled and put up, emphasizing the joint invented by Bogardus; yet nothing is said about the interior construction, although Bogardus' patent was taken out for the "Construction of the Frame, Roof, and Floor of Iron Buildings." Unfortunately, the building at Center and Duane

----

*It is hoped that this building can be incorporated into the future Washington Street redevelopment.

Streets disappeared long ago. Was it an all iron building built according to the patent? At present we do not know.

Bogardus did not stop with patenting a system of construction. Prophetic of the skyscrapers of the future were the shot tower and watch towers he built in New York City. One of the latter remains in Mount Morris Park, a handsome, clearly articulated structure which conveys much of the quality and spirit of the early iron technology.

The present volume as a whole contains valuable source material for the student of this period in the United States. It makes possible a comparison of the careers of two men—Daniel Badger and James Bogardus —who worked in the same street and whose talents in many ways complemented each other. The story is a fascinating one which deserves further exploration.

W.K.S.

# CAST IRON BUILDINGS:
## Their Construction and Advantages

*The borders on these pages delineate
the original page size of* Cast Iron Buildings.

# CAST IRON

# BUILDINGS:

## THEIR

## CONSTRUCTION AND ADVANTAGES.

BY

## JAMES BOGARDUS, C. E.

ARCHITECT IN IRON,

IRON BUILDING, CORNER OF CENTRE AND DUANE STS.

## NEW-YORK.

J. W. HARRISON, PRINTER, 447 BROOME-STREET, NEAR BROADWAY.

1856.

## TO THE READER.

———

    I am indebted for this pamphlet to my friend, Mr. John W. Thomson, A. M.   I mention this, not only in apology for the manner in which my name is introduced, but in justice to him as the author.   I endorse every word of its contents.

JAMES BOGARDUS.

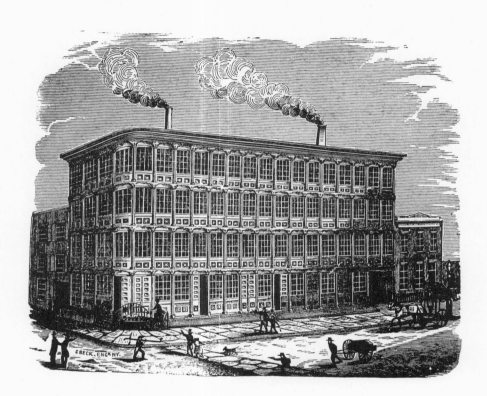

*This plate represents the Factory of James Bogardus, inventor and patentee of Cast Iron Build-ings. It is situated at the corner of Centre and Duane Streets, New York, and is the first cast iron house ever erected.*

This plate represents one of Bogardus's cast iron buildings, with the greater part of its iron work removed, or supposed to be destroyed by violence; in which demolished condition it will yet remain firm. It is designed to illustrate the strength, stability, and safety, obtained by Mr. Bogardus's method of construction; and also the security against an imperfect foundation: advantages possessed by no other buildings.

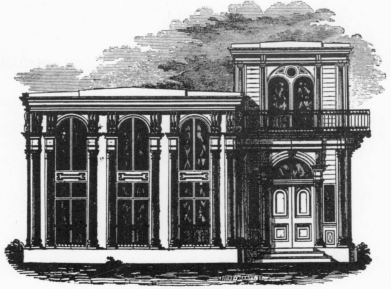

Design for a cast iron Chapel, by James Bogardus.

*View of the cast iron building of Messrs. Harper & Brothers, Publishers, Franklin Square, Pearl Street, New York.*

FORBES.                 TENEYCK Sc.

*This plate is designed to illustrate the facilities of Cast Iron for ornamental purposes. It represents an elaborately finished Capital, with richly carved base of correspondent beauty and workmanship: far too costly for marble; but which, once executed for a pattern, may be rapidly, and cheaply reproduced in iron, with the greatest perfection; and will retain, for ages, its original sharpness of outline.*

# CAST IRON BUILDINGS.

———

So much that is erroneous has been said and written concerning Iron Architecture, and so little that is authentic, that but few are yet acquainted with either its merits or its history. To furnish correct information on this important subject is the object of this pamphlet.

The first complete cast-iron edifice ever erected in America, or in the world, was that of the inventor, James Bogardus—being his manufactory on the corner of Centre and Duane streets, New-York. Its foundation was laid in May, 1848: but a cast-iron model of it had been freely exhibited to visitors at his factory, since the summer of 1847. Previous to this period, the opinion of most men of scientific reputation, was unfavorable to its use for this purpose; and, amongst all classes, there was also a very strong and general prejudice against it. Some accidents had happened from the breaking of cast-iron beams in England, and their cause was ignorantly attributed to the material employed; whereas, the fault lay in the want of proper knowledge and skill in constructing them. These accidents, however, helped to swell the popular prejudice against cast-iron as a material for buildings, and they were frequently quoted in opposition to the inventor; until, by the erection of the building already mentioned, he demonstrated their untrustworthy character. Since that period, he has erected many structures of the same description, not only in New-York, but also in various other cities, such as Philadelphia, Baltimore, Washington, and San

4

Francisco; and persons capable of passing judgment on their merits, have, after a careful investigation, been profoundly convinced of their superiority — that they alone embrace the true principles of safety, durability, and economy. And the inventor himself firmly believes, that, were the public fully aware of its great advantages, cast-iron would be employed, for superior buildings, in every case, in preference to granite, marble, freestone, or brick.

It was whilst in Italy, contemplating there the rich architectural designs of antiquity, that Mr. Bogardus first conceived the idea of emulating them in modern times, by the aid of cast-iron. This was in the year 1840; and, during his subsequent travels in Europe, he held it constantly in view; and cherished it the more carefully, as he became convinced, by inquiry and personal observation, not only that the idea was original with himself, but that he might thereby become the means of greatly adding to our national wealth, and of establishing a new, a valuable, and a permanent branch of industry.

It is impossible for the reader to realize to their full extent, the difficulties which Mr. Bogardus had to encounter in the erection of his first building. Whilst burdened with the care of his factory, and with limited means at his command, he had not only to superintend every detail of its construction, but to hear and answer daily, the same predictions of failure. One would not live in it, if he had it as a gift, for fear lest it would crush itself by its own weight another would not, for fear of lightning: a third was sure that it was not perpendicular, and that sooner or later it would topple to the ground; and a fourth foretold, that if a fire should happen, it would melt the columns, and the whole would fall with one tremendous crash. Others declared, as the universal voice of science, that, in consequence of the expansion and contraction of the metal, it contained within itself the elements of early and rapid decay; and some even asserted that the experiment had already been made in England; that its disastrous failure had been attended with a great public calamity; and that, in consequence thereof, an Act of Parliament was actually then in force, forbidding the erection of cast-iron buildings. But these and other objections, Mr. B. had already thoroughly considered, and found them either to be groundless altogether, or to involve only such difficulties as might be obviated by mechanical means. Meantime, the work on the building, which had been steadily progressing, though but slowly, wholly ceased for a time. This was to many a convincing proof that something was wrong; and the poor unfinished skeleton was christened Bogardus's folly — not knowing that, in the interval, he had commenced and finished the fronts of several stores, on the same pattern, at the corner of Washington and Murray-streets.

It may be also added, that complaints were made to the City Authorities against it, and that some of the tenants of neighboring buildings left their houses through fear of danger: this created some delay, until informed by the chief engineer of the fire department, that the committee had made a favorable report.

This first cast-iron building, Mr. Bogardus's present factory, is of five stories, and was designed to be a model of its kind. Since its erection, it has not been difficult to convince any one who will take the trouble to examine it, that SUCH BUILDINGS COMBINE UNEQUALLED ADVANTAGES OF ORNAMENT, STRENGTH, DURABILITY, AND ECONOMY; WHILST THEY ARE, AT THE SAME TIME, ABSOLUTELY SECURE AGAINST DANGER FROM FIRE, LIGHTNING, AND AN IMPERFECT FOUNDATION.

Whatever be the advantages of cast-iron as a building material, they would be all unavailable, were they not accompanied with stability of structure. But, simple as this problem now appears to be, had it not been hitherto esteemed impracticable, it would not have been left for Mr. Bogardus to solve it. As it is on this point, mainly, that his merit as its inventor depends, a short description is subjoined; and the reader should remember, that the greater the simplicity of an invention, it is the more meritorious.

The cast-iron frame of the building rests upon sills which are cast in sections of any required length. These sills, by the aid of the planing machine, are made of equal thickness, so as not to admit of any variation throughout the whole: they are laid upon a stone foundation, and are fastened together with bolts. On the joints of the sills stand the columns or pilasters, all exactly equal in height, and having both their ends faced in a turning lathe so as to make them perfectly plane and parallel; and each column is firmly bolted to the ends of the two adjacent sills on which it rests. These columns support another series of sills, fascias, or cornices, in sections, of the same length as the former, but of greater height according to the design of the architect: they are separately made of equal dimensions by the planing machine, and are bolted to the columns, and to each other, in the same manner as before. On these again stands another row of columns, and on these columns rests another series of fascias or cornices; and so on, continually, for any required number of stories. The spaces between the columns are filled up with windows, doors, and pannels, which may be ornamented to suit any taste.

It may be here remarked that, in certain cases, the first layer of sills may be dispensed with altogether; and also that, immediately before uniting the pieces, it is the practice of Mr. Bogardus to apply a coating of paint to those parts which are designed to be in contact with others; thus rendering the joints absolutely air-tight.

From this description it is plain, that the separate parts

are so united as to form one stable whole, equivalent in strength to a single piece of cast-iron. Hence, such a structure must be far more firm and solid than one composed of numerous parts, united only by a feeble bond of mortar. On this account it may be raised to a height vastly greater than by any other known means, without impairing its stability in the least; and, were all the columns of any story removed or destroyed by violence, except the four corner ones, or others equivalent in position, the building would still remain firm as an arch; and the greater its height, the firmer it would be.

It is also plain that such a building may be erected with extraordinary facility, and at all seasons of the year. No plumb is needed; no square, no level. As fast as the pieces can be handled, they may be adjusted and secured by the most ignorant workman: the building cannot fail to be perpendicular and firm. Wedges, mortices, and chairs, are all ignored: they are the subsequent inventions of interested individuals, in order to evade the patent; and to render less dangerous, or less apparent, their imperfect and unstable joints. Strength is secured in the simplest and surest way, and at the least possible expense.

It also follows that, a building once erected, it may be taken to pieces with the same facility and despatch, without injuring or destroying any of its parts, and then re-erected elsewhere with the same perfection as at first. The size and form of the pieces greatly favor their portability, which has enabled Mr. Bogardus to construct them in New-York, and export them to distant cities. This quality is of the greatest importance; for it renders every cast-iron building not only a present, but a permanent addition to our national wealth. Who can estimate the annual saving to the city of New-York alone, were all its buildings of this character? The progress of improvement would no longer be accompanied with the work of demolition; instead of destruction, there would be a removal only — a simple change of location.— And to make the calculation properly, we should know not only the present worth of the buildings destroyed, but what was their original cost.

These superior qualities of cast-iron buildings depend mainly upon their mode of structure, without which the rest would be of little avail. We now proceed to consider those superadded advantages which arise more directly from the character of the material employed.

Cast-iron does not indeed possess the character of wrought iron for resisting tensile strain, but it is far superior to it in resisting a crushing force; and it is vastly superior to granite, marble, freestone, or brick, in resisting any kind of force or strain. It may, however, for building purposes, be considered crushing-proof. According to the tables of our best authorities, which have been often verified, a cubic inch of cast-iron can sustain a weight of eighty tons. Now, since a cubic foot weighs four hundred and fifty-five lbs., it follows, by an easy computation, that a column of cast-iron must be ten miles in height, before it will crush itself by its own weight. It will be readily seen that the joint invented by Mr. Bogardus, effectually secures the whole of this important quality; and that thereby he would be enabled to erect a tower or building many times the height of any other edifice in the world, which would be perfectly safe to visitors, in the face of storm or tempest, though they filled it throughout every story, to its utmost capacity.

The great strength of cast-iron, enables us also to enlarge the interior of a house, by lessening the thickness of its walls: a very important item in this city, where ground is of great value.

Cast-iron also possesses the quality of great durability.— Unlike wrought iron and steel, it is not subject to rapid oxydation and decay, by exposure to the atmosphere. And whatever tendency it may have of slowly imbibing oxygen in a moist atmosphere, can easily be prevented by a proper coating of paint, and thus, at a very small expense, be made to endure a thousand years, unaffected by the winds or the weather. On account of this quality, cast-iron houses do not tax their owners with the cost and the trouble of repairs, which are incident to other buildings, in consequence of their perishable character.

Another recommendation of cast-iron is, "its happy adaptability to ornament and decoration." Were a single ornament only required, it might perhaps be executed as cheaply in marble or freestone: but where a multiplicity of the same is needed, they can be cast in iron at an expense not to be named in comparison, even with that of wood; and with this advantage, that they will retain their original fullness and sharpness of outline long after those in stone have decayed and disappeared. Fluted columns and Corinthian capitals, the most elaborate carvings, and the richest designs, which the architect may have dreamed of, but did not dare represent in his plans, may thus be reproduced for little more than the cost of ordinary castings. Ornamental architecture — which, with our limited means, is apt to be tawdry, because incomplete — thus becomes practicable; and its general introduction would greatly tend to elevate the public taste for the beautiful, and to purify and gratify one of the finest qualities of the human mind.

Indeed, so apparent have become the advantages and economy of cast iron for ornamental purposes, that there is danger lest the public overlook the character of the structure of a building, in the contemplation of its architectural beauty; and that, deceived thereby, and by the name of the material employed, they imagine themselves to be admiring a building which has all the superior qualities already described:

whereas, being without stability in the combination of its parts, it is in reality more insecure than our ordinary buildings.

Some have asserted, and it is generally believed, that, as iron is so good a conductor of heat, it would expand and contract so much by the changes of temperature, as to dislocate its joints in a short time, and render the building unsafe. It has been already mentioned, that the assumed impossibility of forming a safe and economical joint for massive structures of cast-iron, was the probable cause why such buildings had not been earlier introduced. It may now be added, that the supposed necessity of making some provision for the expansion and contraction of the metal, was the probable reason why efforts were not made to overcome this difficulty. Indeed, such has been the prevalence of this belief, that, even within a few weeks, a writer in the Daily American Organ, published at Washington, has said:—"Nothing prevents the speedy and general adoption of iron for building purposes, but the practical difficulty in applying it to the substantial portion of architecture, resulting from some of its elementary properties: these are, its expansive and contractable action under the influences of heat and cold, and its extraordinary conducting powers." And in support of his statement, respecting the destructive effects of metallic action, he makes the following quotation from another writer:—

"The first difficulty arising from this source, is the comparatively slight but constantly disorganizing force exerted upon structures of iron or other metals, by expansion from solar heat and contraction by severe cold—a difficulty great in Europe, but much more formidable in this country, where we have such extraordinary extremes of temperature. A distinguished scientific gentleman, speaking of this subject, refers to the monument Colon de la Place Vendome, erected in honor of Napoleon the 1st, and covered with bronze made from captured cannon. 'In this monument,' he says, 'there was experienced much trouble from contraction and expansion. The bronze plates, firmly united by rivets, acted as one stupendous sheet, and buckled under the sun's rays in a most extraordinary manner, acting as a real great pyrometer.'"

If these statements are intended to apply to cast-iron buildings — as they are doubtless meant to do — they are nothing more than the reiteration of all previous writers on the subject, and arise from sheer ignorance: they have been the great bugbear to inventors in this department of art; for, although more or less true in reference to copper, bronze, and certain other metals, they are utterly unfounded when applied to cast-iron. We do not say that cast-iron is without expansibility: we simply assert that the temperature of our climate, throughout its utmost range, from the greatest cold to the greatest heat, exerts upon it no appreciable effect.

A complete proof of this assertion may be had, by examining any of the numerous cast-iron structures, erected by Mr. Bogardus. His factory building has now, for a number of years, been exposed to every change of atmospheric temperature without, and to the heat of steam boilers and the operations of a steam engine and heavy machinery within — and, it should be observed, his engine of twenty-five horse power is placed on the second story, purposely to show the great stability of the building — and yet, so perfect are all its joints, that the blade of a lancet cannot be thrust into one of them; nor can there be discovered, by continual and close observation, where its walls adjoin the neighboring houses, the displacement of a single grain of dust.

The writer of the quotation in the American Organ, continues thus:

"Although ingenious and complicated devices may have partially overcome the effects of expansion arising from this

12

source, they have been wholly inadequate to overcome the much greater expansion from artificial heat in contiguous conflagrations. Iron buildings, as usually constructed, although expressly designed to resist conflagrations in compact cities, have been wholly ineffectual for this purpose. It was found in the great fires at San Francisco that the iron columns and framework of buildings were expanded, and thus warped and thrown out of line, by the heat of fires across the streets, and that the buildings were ruined even before contact of the flames."

From these remarks it may be inferred, that their author either did not know that there was such a thing as cast-iron buildings in existence, or that there was any difference between them and those made of wrought iron. The houses of San Francisco, which are described as shriveling like paper before they came in contact with the flames, were built of sheet-iron, either plain or corrugated; nailed, in most cases, to wooden posts: or, like the better class of English iron houses, riveted to cast-iron columns, and thence ignorantly described as cast-iron buildings.

Cast-iron houses are perfectly fire-proof. Were such a building as Mr. Bogardus's factory filled with the most combustible goods, such as cotton or resin, and the entire interior in flames at once, until the whole was consumed, the building itself would remain firm and unimpaired. Some have said that the columns might melt, and thus precipitate the whole; but this is simply an absurdity, said without reflection: for, it is well known, not only a high and intense heat, but the use of a blast, is required to reduce iron to a molten state; and never yet, in any conflagration, has it been found melted, except in pieces of minute dimensions, and in such situation that the current of the flames created around them an artificial blast. Others compare iron houses to stoves, and tell us that if certain parts be made red hot, and cold water then thrown upon them, it will warp and crack the metal: but this only shows a mechanical defect in their construction; for it is quite possible so to construct a stove that it would stand such a test without damage, though it were repeated many times a day, for years. And that the several parts of Mr. Bogardus's buildings are carefully modeled so as to run no risk of this disaster, he has ascertained by direct experiment for this purpose.

This experiment has been repeatedly verified since in the following way. It has sometimes happened that his columns would be found warped when they came from the foundry. To remedy this defect, he made the column red hot; and whilst in this condition, by means of powerful screws, forced the parts, not only to the position required, but in some cases, as much as six inches beyond it: yet, after cooling, he has invariably found them to be warped, exactly as at first. It may be here added, that, in his later experiments, he endeavored to secure the set of the column, by dashing cold water upon it when red-hot, but equally without success.

It is desirable, in most cases, that the floorings and partitions should be also fire-proof; so that, should a fire occur, it may be confined to the room in which it originated. This may be accomplished by various well known devices, extensively practiced in Europe, but too much neglected in our country. Mr. Bogardus has also devised for this purpose, and secured by letters patent, a plan of iron flooring, to be supported by Cooper's iron beams, or by his own new sectional truss girders, which he is now taking measures to patent, and which may be seen in use in the buildings of Messrs. Harper & Brothers, the well known publishers. These girders, besides having other advantages, can sustain a heavier load than any others of the same weight yet known, and are therefore more economical. And to those who prefer a wooden floor, Mr. Bogardus offers another plan, in which the flooring rests upon the aforesaid truss girders, in combination with a substratum of solid brick-work, and a net-work of iron wire as a substitute for laths.  Of this latter plan a model is now on exhibition at his office.

Cast-iron buildings are also perfectly safe during thunder storms ; no accident from the electric fluid can happen to any person within them.  The metal being a good conductor, and presenting so great a mass to the overcharged clouds, conveys all the electricity silently to the earth, and thus obviates all danger from disruptive discharges.  An iron building, for this reason, requires no lightning rods, because it is a huge conductor itself.  This is a feature deserving consideration ; for many ordinary buildings with rods attached to them, have been struck with lightning, whereby a number of persons have lost their lives; accidents which can never occur in cast-iron buildings.  In them the intensity current is instantly diffused throughout the entire mass, and thus changed into a current of quantity : so that, in any one part, the electricity must be very feeble, and therefore not dangerous to life.

Every style of architecture, and every design the artist can conceive, however plain or however complicated, can be executed exactly in cast-iron; and, in consequence of its having greater strength than any other known building material, it furnishes us with new ideas of the proportional fitness of parts, and thus opens a wide field for new orders of architecture.  Hitherto its use has been confined to factories, stores, lighthouses, and bell towers ; but we hope the day is not distant when we shall see it in our city halls, our state houses, our churches, and their spires.  And Mr. Bogardus himself firmly believes that, had his necessities required him to construct a dwelling house rather than a factory, it would now be as popular for this purpose, as it is for stores.  They would have, moreover, this advantage: being free from damp, they are ready for occupation as soon as finished; nor can they absorb it afterwards, and are consequently, not liable to mildew, and therefore more healthful.

When Mr. Bogardus commenced this business, the use of cast-iron for building purposes, was, in this city, only to be seen in the occasional substitution of a water pipe, or a rude solid pillar, for the ordinary stone posts of the first story.— It needs not be told here how extensively it is now used.— There is scarce a street in our city, and scarce a city in our country, in which are not to be seen either copies or imitations of his beautiful and costly patterns. His mode of forming capitals, a valuable invention which he did not patent, is now in universal use: and, not content with this great and honest addition to their business in the construction of columns for the first story, some have already attempted to evade his patent for house building. As a substitute for his safe and simple joint, wedges, mortices, chairs, and other complicated devices, have actually been patented; and in order to secure their columns, some have fastened them with tie-bars to the wooden girders! Of these contrivances, which are, all of them, mere subterfuges for the evasion of his rights, some are absolutely dangerous; and of the remainder, the best are not only inferior in stability — being liable to dislocation by the displacement of the wedges — but so much more expensive, that the value of the extra iron necessary for their construction, without any regard to the work spent upon it, would alone be sufficient to pay for the cost of erecting a superior building. Far from attempting or desiring to monopolize the business — for the demand promises to be sufficient to support very many large establishments—he is ready to grant the privilege to build, for a fair renumeration, so small as to leave no inducement to infringe his rights as the inventor.

Mr. Bogardus is prepared to carry out designs in cast-iron for public and private buildings of every description, light-houses, towers, &c., and refers to the following gentlemen for whom he has already erected buildings:—

Messrs. HOPKINS & BROTHERS, Barclay-street; Messrs. TATHAM & BROTHERS, Beek-man-street; Dr. J. MILHAU, Broadway; MESSRS. HARPER & BROTHERS, Franklin-Square, Pearl-street; MESSRS. H. SPERRY & Co., Broadway; MESSRS. McKESSON & ROBBINS, Ful-ton-street, New-York. Mr. M. S. SHOEMAKER, of Adams & Co.; MR. A. S. ABLE, Sun Building; MR. E. LARRABEE, Baltimore. Messrs. ADAMS & Co.; MR. F. COYLE; MR. M. SHANKS, Washington, D. C. MR. D. MIXER, Charleston, S. C. Mr. JOHN PARROTT, San Francisco, California, &c., &c., &c.